Venture Capital and the European Biotechnology Industry

Also by William Bains

Biotechnology from A to Z

Venture Capital and the European Biotechnology Industry

William Bains

palgrave
macmillan

Every effort has been made to contact the copyright holders, but if any have been inadvertently overlooked the publishers will be pleased to make the necessary arrangements at the first opportunity.

First published 2009 by
PALGRAVE MACMILLAN

Palgrave Macmillan in the UK is an imprint of Macmillan Publishers Limited, registered in England, company number 785998, of Houndmills, Basingstoke, Hampshire RG21 6XS.

Palgrave Macmillan in the US is a division of St Martin's Press LLC, 175 Fifth Avenue, New York, NY 10010.

Palgrave Macmillan is the global academic imprint of the above companies and has companies and representatives throughout the world.

Palgrave® and Macmillan® are registered trademarks in the United States, the United Kingdom, Europe and other countries.

ISBN-13: 978-0-230-21719-5   hardback
ISBN-10: 0-230-21719-2   hardback

This book is printed on paper suitable for recycling and made from fully managed and sustained forest sources. Logging, pulping and manufacturing processes are expected to conform to the environmental regulations of the country of origin.

A catalogue record for this book is available from the British Library.

Library of Congress Cataloging-in-Publication Data

Bains, William, 1955–
    Venture capital and the European biotechnology industry / William Bains.
        p. cm.
    Includes bibliographical references and index.
    ISBN 978-0-230-21719-5 (alk. paper)
    1. Biotechnology industries—Europe—Finance.   2. Venture capital.
    I. Title.

    HD9999.B443E85125   2009
    660.6068'1—dc22   2008029943

10   9   8   7   6   5   4   3   2   1
18   17   16   15   14   13   12   11   10   09

Printed and bound in Great Britain by
CPI Antony Rowe, Chippenham and Eastbourne

Contents

List of Tables   vi

List of Figures   vii

1 Genes and Money in Europe: Why Did It Fail?   1

2 The Biotechnology Industry and Venture Capital   5

3 Europe Falls Behind   23

4 Why Does Europe Do So Poorly? Some Inadequate Explanations   42

5 Underinvestment in European Biotechnology   55

6 Public Markets and Underinvestment   73

7 VC Management of Companies   86

8 VC Effects on Business Efficacy   109

9 Investor Blockade of Business   123

10 Pomp and Circumstance   140

11 The Real VC Business Model   150

12 What to Do about It: Government   173

13 What to Do about It: Business   189

Notes   198

References   201

Company Index   216

Topical Index   218

Tables

2.1	Comparative market capitalisations	20
3.1	VC and technology company formation activity in Canada	28
3.2	VC capacity versus population	29
4.1	Where biotechs collaborate	51
4.2	Government spend on health R&D	53
5.1	Factors leading to company failure	56
7.1	NVT stay and company success	105
11.1	Compensation package for top US banks	155
12.1	Company change as a result of innovation	176

Figures

2.1 The money cascade for VC   11
2.2 Numbers of biotech companies   15
2.3 Market capitalization of public biotech companies   16
2.4 Total employed by the biotechnology industry, US and Europe   16
2.5 Employment gap between the US and Europe   17
2.6 Cumulative drug approvals from biotech companies, 1980–2005   18
2.7 Market capitalization of public biotechnology companies, end of 2005   19
2.8 Market capitalization excluding 'old giants'   20
2.9 Biotech aggregate capitalisation versus country size   21
3.1 Rate of formation of early biotech start-ups   24
3.2 VC returns from European biotech VC funds   31
3.3 Simplified fund life cycle example   33
3.4 NASDAQ comparators for VC returns   36
3.5 Scientist career total earnings from different approaches to science commercialisation   39
4.1 Business models for biotech companies in 1988   43
4.2 Survival of Cambridge area biotech companies   44
4.3 Date of foundation of 1998-vintage companies   45
4.4 Employment mobility in European biotech companies   48
4.5 Expatriate experience of European CXOs   49
4.6 Scientific publications by country   52
4.7 Origins of start-up companies   53
5.1 Amount invested in biotech by company age   57
5.2 Investment in UK and US biotech companies   58
5.3 Pharmaceutical company patent filings   60
5.4 Failure in UK biotechnology start-ups   64
5.5 Time taken to invest in UK biotech companies   65
5.6 Technical due diligence levels in start-up biotech fundings   67
6.1 Average pre-IPO valuation and average cash raised, US vs EU   75
6.2 US and European biotech IPOs 1995–2002   77
6.3 Investment versus deals for genomics companies   78
6.4 Time to IPO for European and US biotech companies   80
6.5 Ratio public to private sources of finance   83

7.1 How long VCs look at different aspects of propositions 89
7.2 Reasons for VC rejection of business plans (UK) 90
7.3 Shareholder veto powers 92
7.4 European VC claims for non-financial support
for companies 95
7.5 Impact and cost of advice 96
7.6 Fraction of VC time spent in different activities 98
7.7 Investments per partner per year 99
7.8 How many Boards do VC investors sit on? 100
7.9 Survival of NVT after foundation 103
7.10 X-axis – share value today as a fraction of share value
at IPO 107
8.1 Business sector of public biotech companies (2004) 110
8.2 Business sector of private biotech companies (2004) 111
8.3 Industry areas of research-based biotech companies
in 1988 112
8.4 The medicines development process 113
8.5 Geron stock price 116
9.1 Preference share effects 124
9.2 Preference multiples in 44 UK biotech investments,
1998–2005 126
9.3 Causes and effects of Pref shares 127
9.4 Arakis exit outcomes for major share classes 133
9.5 Shareholder commonality in merged companies 135
9.6 Commercial outcomes of mergers 136
10.1 Big company experience and biotech success 143
10.2 Where do Pharma-experienced biotech execs come from 144
10.3 Management techniques used by SME technology
companies 148
11.1 VC yield from management and Carried Interest 154
11.2 Time gap between fund closings 157
11.3 VC funds raised versus VC-backed IPOs 158
11.4 Average size of specialist VC funds in Bio- and
general tech 168
11.5 Fund increase versus inter-fund gap 169
12.1 Unemployment versus small firm employment 178
12.2 Company creation versus national initiatives 180
13.1 Patent publications from some UK Universities 194

1
Genes and Money in Europe: Why Did It Fail?

This book is slightly unusual. Most business books are about how to run a successful business. This one aims to be a detailed, fact-based primer on how to fail in one of the world's most exciting and fast-moving industrial sector – biotechnology.

The biotechnology industry is replete with good-news stories, icons of wealth creation, morale-boosting tales of corporate heroics. Many reviews of 'how to make a successful biotech company' explicitly ignore the example of failures, which is rather like making a film called 'how to live forever' in a graveyard.[1] There are many hard-working, expert and dedicated people working in biotechnology, in science, in venture capital (VC) and in the companies which bring them together. And they have created a few outstanding successes. But any new business endeavour is an experiment. All Western economies are constantly trying out new types of business, creating what Metcalfe calls 'restless capitalism' [1] where knowledge of past events drives not just new products but new ways of doing business and finance, and the knowledge thereby gained drives yet more innovation. The biotechnology industry was such an experiment, and by the measures I present here, the late twentieth century European biotechnology experiment failed. This is the overwhelming reality of the industry in Europe, and especially in the UK, recounted by nearly every biotech entrepreneur I have talked to in two decades of doing business in biotech. On the European side of the Atlantic, 'biotechnology' is a byword for opportunities wasted, careers destroyed and private lives ruined, of waste of government and personal resources, lost wealth and opportunity for European economies and of throwing away biomedical opportunities that could be of real, lasting value to us all.

None of us are getting younger, fitter or healthier. Throwing away the opportunity to do something about that, for short-term financial gain, is not a great idea. Throwing it away in order to make a loss seems positively absurd. So I have spent five years researching why this is happening, and this book is the result.

It is the contention of this book that there are three principle contributors to this failure: the management, the government and the investors. The manifold failings of greedy entrepreneurs, egomaniac founding technologists, power-crazed CEOs and academics completely divorced from reality, are covered in many other places. This book does not wish to claim that these are not real problems: a business must be built on good commerce, good management and a practical growth strategy, not just 'great science'. We will cover how realistic the views of company creators and management are in Chapter 7. Less often critiqued is government policy across Europe. To summarise policy to the point of parody, it has been this: some enormously successful US biotech companies started as start-up companies based on cutting edge university research, so if we are to stimulate the creation of spin-out companies based on University research then they will become huge and successful biotech companies. The flaws in this Cargo Cult Economics are starting to be understood, but the damaging knock-on effects are generally not, so I will analyse this cause of failure in Chapter 12.

But these factors would not have resulted in an industry that is a Bonsai[2] imitation of its grown-up US counterpart if not for the influence and effect of the most powerful force in the economics of biotechnology. That is, the private investment industry, called (rather inaccurately in Europe) VC.

The assumption built into most analysis of VC in biotechnology is that it is a 'good thing'. This image of VC is rather different from the experience of many companies and entrepreneurs in European biotechnology. VC is often, and loudly criticised by entrepreneurs. Entrepreneurs often feel that VC feeds on their efforts and returns nothing, a feeling reflected in the well-known joke that 'VC' stands not for 'VC' but 'Vulture Capital'. (But vultures are useful animals, recycling the remains of carcasses that have already fed the lions or wolves. I have long thought that the feelings of most entrepreneurs are better captured by the phrase 'Vampire Capital'. The vampire stalks a beautiful, defenceless young thing with 50 years of success and happiness ahead of her, seduces her with illusions and then sucks the life out of her[3]) But this type of abuse, even though derived from deep experience of the problems, is usually written off as sour grapes from entrepreneurs whose dreams of wealth and prestige have not been realised. This literature, large and largely web-based, is not a terribly reliable source of statistical evidence for economics. To counter it, VCs and their spokes-organizations such as EVCA[4] and BIA[5] provide case studies of how VCs have rescued ailing companies from the grasp of their incompetent founders and turned them into commercial and technical successes – again, we should view such anecdotal data as suspect.

This book, therefore, aims to provide a detailed, factual analysis of the effect that the present model of investment has had on high-tech start-up companies in Europe, and some analysis of the mechanisms (and hence reasons) behind it. It is the factual backup for what many entrepreneurial scientists and businessmen in the biotechnology community have learned as unproven but obvious over the last decade, and as such I hope it is some use to them. I hope that many readers, even if they do not agree with my conclusions, can use this source of factual data about the industry and its investors for their own purposes.

I have not tried to propose what one entrepreneur colourfully called 'A universal theory of VC suckage' [2]. I do not believe there is one. Rather, investor and entrepreneur activity, driven by a range of *ad hoc* reasons, and the VC business model which propels these which I describe in Chapter 11 , is a rational response to the economic environment in which VC management groups find themselves. There is no plot to do down entrepreneurs, and make sure that innovative drugs never reach the market – it is a mistake to see incompetence as conspiracy. Nor do I claim that 'sorting out' VC would make European biotech rise, phoenix-like to out-compete the Americas. This is an analysis of one aspect only, and VC apologists remain free to dismiss what I present here as one small aspect of the problem, possibly true but of no significance. That is their prerogative.

I have also not tried to tie this to formal economic or business theory, although I have referenced some of it where it is relevant. There is a mass of literature on what *should* happen in high-tech company finance, based on logical and mathematically sound economic arguments, (which usually assume that people are informed, rational agents, which undermines their arguments from the start). There is a mass of literature (mostly PR material for various interest groups) on what people *would like* to happen. This book, however, is based on data of what actually *did* happen. There is no point, therefore, in contesting my assumptions: my assumptions are *what actually happened.* You may choose to contest my choice of facts and my conclusions. But saying 'that just is not true so' will not refute the argument.

Having said that this is a book based on fact and statistics, I do not pretend that I started out with no idea of what the conclusion would be. It is clear that VC practices in Europe have at very least been ineffective at creating successful companies, at worst been positively toxic to entrepreneurship and company creation. What I have tried to do here is demonstrate that the intuitive feeling that this is so, felt by many entrepreneurs, managers, personal investors and not a few VCs, is backed by the facts. Since the start of 2008, the credit crunch, and the flight of VC from funding biotech of any sort in Europe, such views have become more common, and indeed some will regard this book as a post-mortem of a business model already dead. But the same things were said in 2002, and VC survived, and indeed flourished in its own way. The VC model is not dead – at most, it is hibernating through our current economic winter. This will be an enduring issue, and it is my hope that this book helps address it.

Many people have helped me in writing this book, and in the research over the last few years which I summarise here. My sincere thanks to them all. Quite a few wanted to stay anonymous, for reasons that will become obvious, so I will not name names. Several of them advised me not to write it at all if I ever want to work in a VC-backed industry again. Ladies and gentlemen, I will take that risk, again for reasons that I think this book will make clear. My thanks also to the staff and students of the Cambridge University MPhil in Bioscience Enterprise course, to the good folk at Palgrave Macmillan for taking a risk with me, and steering me clear of some even greater ones, and to you, gentle reader, for reading at least this far. Now read on!

2

The Biotechnology Industry and Venture Capital

This is not a book about the *technology* of biotechnology, but about the *business* of biotechnology. The latter is dependent on the former, but can be described without ever explaining what a gene is.[1] I will try to keep it that way.

The biotechnology industry takes novel life science discoveries or technologies and turns them into products [3]. It is one of the highest-profile science-based industries, with over 4000 companies, $63 billion in revenues and $20 billion in R&D investment worldwide, with a quarter of those companies and three-quarters of the revenue in the US [7]. The highest profile of such companies are the 'biotech start-ups', new companies created to commercialise specific pieces of technology, and these are often considered typical of high-tech entrepreneurship as a whole. These companies are the subject of this book.

The modern biotechnology industry started in the mid-1970s, when a number of pioneers realised that the new advances in explaining how living things worked in terms of molecules could be put to practical use. The initial applications were in the US, with the creation of companies such as Cetus, Chiron and Genentech. But Europe was not far behind, with the foundation of Celltech and Biogen and the growth of companies such as Serono (founded in 1906 but revitalised as a modern biotech company when it moved to Geneva in 1977 and launched its growth hormone product). In the US and Europe at this time several larger companies, notably Kabi, Eli Lilly and Novo, were also active in building a biotechnology research infrastructure.

However, what distinguished the new industry from Eli Lilly, Novo and Serono was the new companies created explicitly to exploit the application of the 'new biology'. It was the creation of an entity called a 'biotechnology company' in the late 1970s in the US and early 1980s in

Europe that galvanised scientist and government alike. Scientists saw a way of sidestepping the size and stratification of major corporations and the University system and turning *their* science into money and fame. Governments saw a way of creating a whole new, clean, knowledge-driven industry from their investment in further education and basic research. And investors saw products that would revolutionise the world and make them rich in the process. In 1980, when Genentech floated and its stock price doubled in the same day, this seemed an industry that could not fail. Move over Microsoft, the genes were coming.

The only drawback, and the major difference between Microsoft in 1973 and Cetus in 1973, were that biotechnology required substantial investment to produce a product [8]. While a teenager could (in 1973) produce a commercial computer program in his bedroom, biotechnology took laboratories, PhD trained scientists and many years: Amgen took two years just to isolate the gene for erythropoietin (EPO), its initial best-selling product, which had to be followed by five more years of scale-up, production, trials in animals and humans and then approval by the FDA (US Food and Drug Administration) before they could generate revenue: by today's tightened regulatory standards, that is considered exceptionally fast. So the new companies were dependent from the outset on people to provide investment capital to support their very expensive R&D efforts.

Of course, this only applies to companies that plan to do R&D. From before 1980 there was an assumption in the mind of scientists, investors and journalists that 'a biotechnology company' was one doing research to create new products (and we will later examine why that might be). But it is quite possible to set up a 'biotechnology' company that just sells the product of existing technology, and if this is not to a highly regulated market such as health care, then they can do that almost immediately. So there have been three types of biotechnology companies. Small 'garage biotechnology start-ups' did not attract major market attention until, like Invitrogen, they floated on the basis of being a profitable business. These companies are often characterised as 'lifestyle businesses' set up to generate cash to sustain what their founders are good at. It is a business model sneered at by VC, but it is capable of building multinational corporations (Microsoft and Google started as teenage lifestyle businesses). A textbook example in biotech might be Bangs Laboratories, founded by Leigh and Sonia Bangs in 1988, literally in their garage, to produce specialised reagents for research scientists, or UK companies such as Oxford Ancestors or AbCam (the latter floated in 2005). If they receive investment at all, it is from private investors – 'Business Angels'.

Exact figures on Angel investments are hard to come by, but it is estimated that even in the UK, where there is a relative lack of wealthy individuals and a culture that runs counter to private risk investment, Angels probably invest as much as VC and, because they invest in smaller companies, actually support eight times as many start-ups [9]. We will address the relationship between VCs and Angels at the end of the book.

In fact, many of today's European success stories started this way. Cambridge Antibody Technology had no venture investment – it was funded by research programme and corporate investment from its partners, notably Peptech. So the second business model, widely ignored but proven to be successful, was the 'big garage' start-up, the company which started with a vision to be a major corporation but which funded itself from non-equity income from the start. The evidence suggests that this is in fact the 'normal' model for company development in Europe, with over 50% of the publicly quoted UK biotechnology companies in 2006 having never received any VC investment at all.

However, the most widely recognised third business model for creating and building a biotech company is to start with some science and an idea of the product it can be turned into, gather investment to realise that product, spend years researching it and only then sell the product to generate revenue. The usual product is a new drug, for a variety of (partly fallacious) reasons that I will discuss later. Such ventures are enormously risky, with drug discovery and development typically taking 12 years and costing $350 million even if successful [10]. This type of technical development costs too much to be sustained by personal investment, so the biotechnology industry has relied on investment from institutional investors who are able to put $10 million or more into a company at each investment stage [11]. This is the role of the VC group.

VC groups in the US are institutional investors who specialise in high-risk, high-return investments where their expertise and involvement in their investee companies can add value that others cannot see or realise [12]. Banks and pension funds will not commit funds to a project based on science that they do not understand with a 50% chance of complete failure, and only a 2% chance of major success. VC can, and sometimes does.

In addition, companies requiring smaller amounts of investment might turn to VC because it is the only source of investment that they know. Many entrepreneurs comment that closing an investment deal with a Business Angel is not hard (in the sense of being excessively drawn-out or punitively dilutive), but because Angels do not generally advertise their wealth, finding them in the first place is the major barrier [13]. Indeed,

Angels often have a 'passion for secrecy', and so entrepreneurs do not realise the volumes of cash available [14]. In the 1990s, many biotechnology companies may have gone to VC simply through ignorance that other alternatives existed. The rise of Angel Networks post 2000 is changing this landscape.

Because of this close link between VC and biotech, the industry is often seen as being almost symbiotic with VC in the US and in Europe [15–17]. Kenny comments, 'Biotechnology has emerged as an industry largely because of one economic institution: VC.' [18], page 133. Many spin-out companies aim to obtain VC as an integral part of their development.

Although this explanation is only partly correct, as we shall see, it is certainly true that in countries such as Japan, with a very high scientific reputation and output but minimal VC activity, there was little biotechnology industry in the Western sense in the 1980s and 1990s.

I should digress here briefly into the mechanics of VC investment in any company, including biotechnology companies, to make sure the reader is familiar with the processes and jargon that we will be handling later in the book. Readers familiar with the VC investment process may skip ahead to page 9.

Companies wishing to raise funds from investment do so by issuing new shares or 'stock' in the company. This is distinct from existing shareholders selling their shares – in the latter case the shareholders get the cash, the company does not, so that does not help raise funds for the company's activities. So the company creates, or 'issues', new shares to sell to the investor, and the company itself gets cash as a result. The obvious question is how many shares a company must issue to get a specific amount of cash, and that is defined by what fraction of the company those shares represent and what the value of the company as a whole is. The former is a minor technical matter, but the latter is critical, and is the issue of *valuation* of the company.

A company's valuation depends on its current assets and likely future assets, including revenue (an important point we will return to in Chapter 8). Obviously, these will change with time. Discovering what a company's probable future asset value is likely to be is hard enough for a company with a ten-year trading record. When the company's sole present asset is a research programme and some patents, working out its value in five years' time is almost impossible. So it takes a lot of work to evaluate a company's value, and this is not something the VC wishes to do continuously. For this reason, and because of the complexity of the investment mechanisms (some of which we will discuss further in Chapter 7), investors make investments at distinct times. So an investment is made as a chunk of cash invested on specific terms (often split into *tranches* so that the investor can keep it in their bank account and earn the interest rather than have the company take the interest). Once that investment is spent, the company might need to ask for more cash, and so go round the process of seeking investors, negotiating terms including the valuation and closing the legal documentation all over again: a 'unit' of investment made in this way is usually, therefore, called an investment Round.

Usually an investment round will have more than one investor: a syndicate of investors will act together to reduce their mutual risk. One of them will 'lead' the round, taking the cost of drafting the legal paperwork and performing 'due diligence' on the company (i.e. examining its technical, commercial and financial fitness) – the lead investor is not usually out of pocket, though, as they inevitably recharge those costs back to the company once they have invested. Rounds can be 'major', that is, a substantial amount of investment made by a syndicate including at least some new investors, a 'closed' round (in which some members of the existing syndicate put in some more cash, but no new investors are invited to join), or a 'top-up' round where some or all the existing investors put in a small amount of cash to keep the company going. Other concepts, such as share classes and 'investor protection' provisions will be discussed in Chapter 9.

The concept of a symbiosis between VC and entrepreneur in biotechnology extends beyond one supplying money and the other spending it. In the US, biotechnology company creation is often a joint activity between VC and scientist (e.g. [19]). Genentech was famously started by an alliance between a venture capitalist and a scientist: the two designed the company together from scratch. The model in which venture capitalists, business professionals and scientists come together to create new companies has remained alive and well in the US. In Europe, however, the idea that a venture capitalist should be involved in the creation of companies is much rarer. In the last 20 years, only Merlin Ventures (as it then was), and Avlar Biosciences in the UK, Technostart in Germany and Abingworth across Europe have made a significant public play of their intention to actually create new companies, as opposed to investing in opportunities that were brought to them. Both Merlin and Technostart effectively ceased to do this in 1999–2000, switching to a more conservative role of investing in pre-prepared propositions (Merlin changed its name to Merlin Biosciences to signal this change in strategy). A lot of European VCs say that they play an active role in supporting and growing the companies in which they invest, a claim we will test later, but they do not claim to create companies.

This reliance on VC posed a dilemma. VC requires returns of 20%–30% annualised on its investment portfolio: if half its portfolio is going to fail (a not unreasonable assumption, as we shall see), that means its 'stars' must be able to generate 60% annual returns or more. As you do not know which are the stars and which the meteorites, every investment should be capable at the start of generating 50% annual return on investment (ROI) or more.

This is not greed (or anyway, not purely greed), despite what entrepreneurs believe. Figure 2.1 illustrates the problem. The only actual source of funds for investment in biotechnology companies is the economic activity of the country, that is, people who do work. This is put into banks and pension funds, which is invested in funds-of-funds and other investment vehicles, which in turn invest in VC funds, which invest in companies. Venture funds are typically fixed-length funds – they raise capital, identify investment opportunities, invest in them, manage those investments for a time and then sell before the end of the life (typically ten years) of the fund.

Each of the parties involved in this funding cascade requires a management fee, and most expect a profit cut at the end which must be generated within the lifetime of the fund. If the general public expects a better-than-inflation return for its investment (e.g. around 6%), the management fee and profit cut means that for every £100 the public provides only £60 gets invested in a company at the right-hand end of the diagram, and the timing of the fund cycle means that it has to turn this into £220 in a period of 5–6 years in order to provide the £179 to hand back to the general public to generate a 6% rate of return. This means the company must grow at a compound growth rate of 30%. This assumes that every investment is successful – if it is assumed that half the investments will fail to make a significant return, then the rest must be capable of 60% growth, and so on. Thus, typical ROIs that VCs required their seed stage investments to be capable of achieving are for the UK 30–55%, 36–55% for France, 31–35% for Benelux in the late 1990s [20].

The problem with this is that no company can sustain 60% ROI for decades: the logical limits to growth mean that it has to slow down at some stage. The specific application to the biotechnology industry can be calculated from the known costs, time and failure rates of the drug discovery business, and I do this in more detail in Chapter 8, but the answer is that VC investing in drug discovery cannot possibly gain the ROI it wants from product sales. In any case, if investors are investing

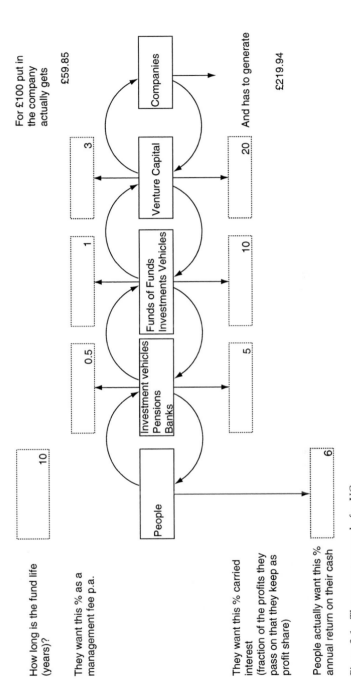

Figure 2.1 The money cascade for VC
Source: Graphic from author's model of the 'money filter' applied to VC.

from a fund which must realise its assets and pay back *its* investors within ten years, it will be incompatible with a product development process that requires 12 years to create a product.

The solution to this conundrum is of course to pass the investment on to others with a lower tolerance of risk but a lower expectation of return at a mid-stage in the development process. This is what the VC-backed biotechnology industry does: companies are created, funded by VC and then (if successful) floated on a public stock exchange or sold to other companies, to provide both a return for their VC investors and a new shareholder base with lower risk acceptance but lower expectations of growth. A typical flotation point is when the company has achieved some evidence that its product will work and be profitable, for example, in the case of a company discovering new drugs, when some early clinical trials have been completed successfully. UK stock exchange rules were changed in the 1990s explicitly to allow loss-making companies to float when they had clinical trial results just so as to facilitate this business model.

However, this means that the biotechnology industry is primarily an *investment* industry, not a science and technology industry or a product industry. The 'product' of early stage investment in biotechnology is not a successful drug, but a floatable company. As Murray [21] says, for all VCs 'Investee company clients should more properly be seen as the "raw material" or "firm stock" from which the investors' value added and capital gain is derived.' 'Development' means obtaining investment and then getting the company ready to float, not doing science. The principal concerns of new companies in the first few years are raising money from investors and then providing those investors with an exit, usually through flotation, sometimes through trade sale [22].

Of course, if the science fails, or looks like it is going to fail, or even looks similar to other science that is going to fail, then public market investors will not buy the shares, it will not be possible to float and the business will fail to achieve its goals. Science and technology remain critical. But in the VC-investment-driven business, the primary object of that science and technology is to gather investors, and only secondarily is it to make products. Thus, the science, management structure, product selection, development programmes and even the location of premises and choice of name and logo are geared towards these investment goals. This is the real nature of the 'synergy' or 'symbiosis' between the VC industry and biotechnology. It is not merely that the one needs the other. The requirement for VC cash and the demands of the VC business model shape everything the modern biotechnology industry is. It is a principal contention of this book that this is a bad thing.

Despite this apparent contradiction between the aims of biotechnology companies (to build sustainable businesses) and of the investment groups that run them (to build investment portfolios by selling businesses), the biotechnology industry has grown to a multibillion-dollar concern, mostly in the US, launching new drugs every year. Its role-model leaders, Genentech, Amgen, Genzyme and Biogen Idec, are now ranked not as start-up companies but as mid-sized pharmaceutical companies, based on market capitalisation. VC claims that it has been the principle behind this, in the US and in Europe.

Success, it is often claimed, has many fathers; failure is an orphan. It is hard to get an objective view of where success comes from. During discussions in 2006 I met three VC groups who claimed that their work had created the success in Solexa. (My view, that the scientists who had worked 12 hours a day for five years to get the technology to work might have had a hand in it, was met with polite bafflement.) Merlin and the EIF (European Investment Fund) [23] both claim Ark Therapeutics' success as 'theirs'. Key executives at both Merlin and Avlar offer as proof of their competence that they created Chiroscience[2]. The reality of course is that it takes many people to create a business, and all can justifiably claim to have made a difference. Oddly, though, when a company fails it is all the executives' fault, not the investors', and as their employment contracts usually (in the UK, anyway) forbid them from saying anything to the contrary, that is the story that gets out. Rather than judge the success of the VC industry in creating the European biotechnology industry by what the VCs themselves claim, I will judge it on the facts, and the first fact to question is whether the European biotechnology industry is actually a success.

Failure, or failure to succeed, in terms of the returns that VCs expect is of two types, which I will deal with separately (although of course they are linked): failure to build an industry (i.e. failure to build successful companies) and failure to make money.

The European biotechnology industry is substantially smaller than its US counterpart on most measures. Only in the number of companies does Europe exceed the US [24], and as the creation of a company as a legal entity requires no significant effort, intellectual property or business activity, it is not clear whether raw company number is a useful measure of industry size or success. The industry is not a failure like Tegenero – a spectacular crash-and-burn followed by liquidation of the company after its only product nearly killed the six people it was given to [25]. Very few biotech companies 'fail' in this sense, for reasons I will discuss in Chapter 11. But has it succeeded as well as it could, or could it have done more?

This is a relevant question. There is a tendency in analyses of the biotechnology industry (or any other new economic area) to observe what has happened and then explain this as the successful result of whatever the postulant was doing at the time. In medical trials, we set out the definition of a successful outcome at the start of the trial, the trial is conducted and then the actual result is compared with the predicted result. We do not (or should not) retrospectively define 'success' as what happened in the trial. Business studies should do the same, and if we apply those criteria to the European biotechnology industry, the trial that has been running for 25 years since 1980 has been a failure.

The 'control arm' of our economic trial, the arm that shows what could have happened, is illustrated by North America. The population sizes of Europe and North America, the number and quality of research centres and the overall economic wealth are roughly the same, to within a factor of two. But the fate of their biotechnology industries has been hugely different: the European biotechnology industry is a bonsai tree of an industry compared with its US counterpart – the same number of companies but tiny in size, bearing miniature technical and commercial fruits of success. So some other factors that are different between the regions must be the cause of this.

The differences in size and maturity are often referred to as the 'European Lag', as the statistics of the European industry today look rather like those of the US industry in the 1980s. The inference is usually made that Europeans only cottoned on to the idea of a biotechnology industry about 20 years after the US. Compelling reasons, based on the differences between European and US financial and entrepreneurial environments, are usually cited for this.

There is no disagreement that Europe has been slow to create new companies in the biotechnology space, only reaching the rate of company creation that the US achieved in the mid-1980s a decade later (Figure 2.2).

European commentators, and especially European politicians, in the late 1990s made much of this curve as showing that the European industry had at last caught up with the US one. Political and hence journalistic energy tends to be focused on start-ups in Europe, following the logic of 'if you build it, they will come'. Universities were judged (and to an extent funded) on the number of spin-off companies they claimed to have created. So all these groups emphasise the curves in Figure 2.2. However, a company as a legal entity is extremely simple to create. It is fallacious to compare a company set-up as a legal 'shell' for £100 with one unpaid director with an Amgen or a Genentech. While start-ups are the seeds of the industry, and without company creation new company

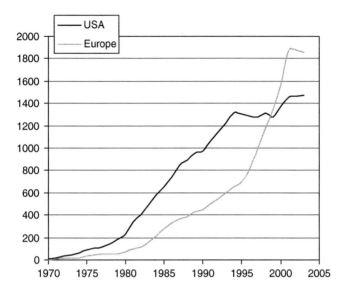

Figure 2.2   Numbers of biotech companies
Source: Data from Ernst and Young reports, Bioscan database.

growth cannot happen, it is a small component of the success of an industry sector as a whole.

Whatever the start-up conditions are, once a company is successfully launched one cannot entirely blame its seed funders, founding management team (most of whom will have departed, as we will discuss later), founding academic science or other initial starting conditions for its lack of success. More relevant are measures of how the company had operated successfully *after* foundation, measures such as the size of the companies concerned and their ability to create products. Growth can mean a wide range of things. Has the European industry grown?

Failure to grow a successful industry is more dramatic, and in this Europe has nowhere near caught up with the US. Market capitalisation of public companies is often cited as a measure, but this has only been systematically followed in both markets since the late 1990s and shows substantial fluctuations due to underlying market variability (Figure 2.3).

An indicator which is less subject to such fluctuations is employment, especially in Europe, where employment law means that hiring new staff has long-term (i.e. at least one month) implications, and so is a reflection of long-term prospects for the company. Comparative employment figures are shown in Figure 2.4.

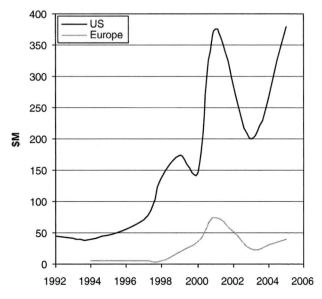

Figure 2.3   Market capitalisation of public biotech companies
Source: Data from [7, 26–30].

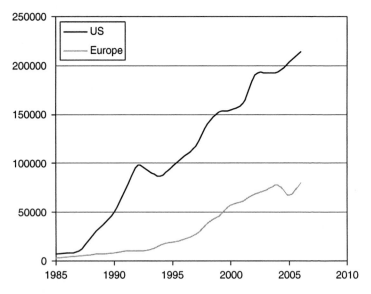

Figure 2.4   Total employed by the biotechnology industry, US and Europe
Source: Data from [7, 26–30].

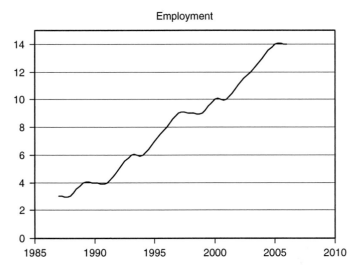

Figure 2.5 Employment gap between the US and Europe
Source: Data from [7, 26–30].

Again a substantial gap between the US and Europe is obvious, but this also points up that the gap is widening. Figure 2.5 analyses the employment data in terms of the gap: for a given date, how many years must we go back in the US employment figures to find the same employment in the industry as at the reference European date? (Thus, for 2002, the European industry was estimated to employ 68,303 people, a size the US industry achieved between 1990 and 1991: the 'gap' is therefore 11 years.)

(The employment growth curves are both approximately linear, so different growth dynamics cannot explain this lag.)

The majority of companies publicly perceived as 'biotechnology companies' work in health care, principally in the discovery of new drugs. If we look at the rate at which new drugs are approved for human use (after a long and very expensive R&D process, which we will discuss further in Chapter 8), we see that Europe is at least 15 years behind the US – the cumulative curve of drug launches today can be thought to be like that of the US companies in about 1990: optimistically, therefore, we could say that Europe is ten years behind the US.

The picture is actually slightly gloomier. In Figure 2.6 European drug approvals have been separated into those launched by two major corporations which have repositioned themselves as biotechnology companies – Serono and Novo Nordisk – and those launched by all

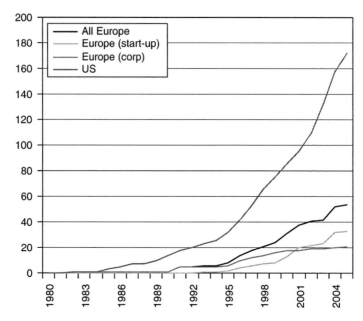

Figure 2.6   Cumulative drug approvals from biotech companies, 1980–2005
Source: Data from the BIO website, Nature Biotechnology [31], Ernst and Young [26].

others. The others are the VC-funded start-ups, analogous to Genentech, Biogen, etc. (In this data set, Biogen count as a US company – they had migrated to the US before they launched a single product.) All the early launches were from this 'corporate' set of companies, established companies that have invested internally in biotechnology. The new start-ups started launching products only in the late 1990s, putting them at least 15 years behind the US.

The nature of the products is also revealing. The first products launched in the US were recombinant insulin, growth hormone, interferon, IL-2 and EPO – extremely high-tech, cutting-edge products for the mid-1980s – and because of their groundbreaking efficacy several became blockbuster sales generators. The first products launched in Europe were a mixture of new variants on EPO, diagnostic reagents and 'me too' drugs (i.e. small technical improvements on well-known agents) like Chiroscience's laevobupivacaine in 1998. Products of this sort can be the basis of good businesses, but they are not the cutting edge of science, medicine or profitability.

For investors, products are secondary to exits, usually seen as the same as IPOs (Initial Public Offering – first floatation of the company shares on a public stock market). Certainly, the largest biotechnology companies have gone public and grown until they are the same size as mid-cap pharmaceutical companies.

The aggregate market capitalisation ('size') of biotechnology companies over time tells us more about the fluctuations of the stock market than about the companies, as we have noted, but a snapshot allows us to compare across countries (Figure 2.7).

The overwhelming dominance of the US is obvious. However, about 45% of this is the 'big 4' – Genentech, Amgen, Genzyme and Biogen Idec – which were founded in the first wave of biotechnology company creation in the period 1975–81. Indeed, the market capitalisations of any of the top five companies are substantially greater than those of the entire UK industry, and Genentech or Amgen could buy every public biotech company in Europe, including Serono, for stock (Table 2.1). (After this table was compiled, at the end of 2006, Serono was acquired by Merck KGaA, removing one of the last major independent companies from the European industry.)

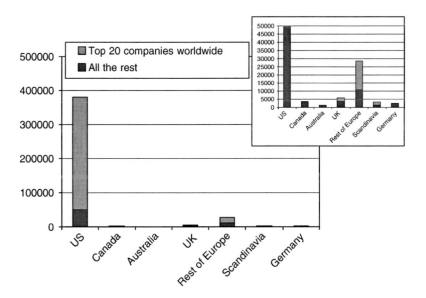

Figure 2.7   Market capitalisation of public biotechnology companies, end of 2005
Note: Where a company has multiple sites, the site of the corporate headquarters is taken as the location of the company. Shown is the capitalisation of all companies and of the top 20 companies worldwide. Insert – same data plotted at 10× expanded Y scale.
Source: Data from [32], company websites.

Table 2.1 Comparative market capitalisations

Company/region	Market capitalisation ($ in million)
Genentech	84,390
Amgen	80,292
Gilead Sciences	29,792
Genzyme	15,571
Biogen Idec	15,370
All European companies	37,973

Note: Top 5 biotechnology companies (worldwide) and all European and all UK companies.
Source: Data from [32], company websites, *Financial Times*. Data for the end of 2005.

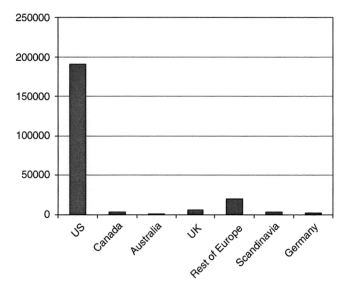

Figure 2.8 Market capitalisation excluding 'old giants'
Source: Data from Figure 2.8 excluding Amgen, Genentech, Biogen, Chiron and Serono.

It can be argued that these companies were created at a unique time, and it is unreasonable to expect others to emulate their experience, just as one cannot found another Microsoft today. This begs two questions. Why did Celltech (founded 1981) not do the same in the UK? And why have Gilead (founded 1987, market capitalization $29.7 billion) and Celgene (founded 1986, market cap $15 billion), outperformed British Biotechnology (founded 1986, market cap as Vernalis $360 million)?

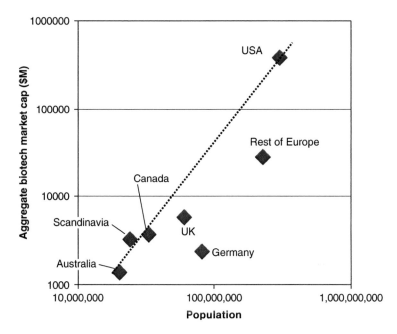

Figure 2.9   Biotech aggregate capitalisation versus country size
Source: Capitalisation data from [32], *Financial Times* and company websites; population data from CIA World Factbook (https://www.cia.gov/library/publications/the-world-factbook/index.html).

But if we accept the uniqueness of the industry in California in the late 1970s, we can exclude these companies. However, to do so we should also exclude Serono, which was founded in 1907, from the European group. This is done in Figure 2.8.

This still shows the US dominance. Interestingly, it also shows that the UK's claim to be the leader of the European industry is questionable. In the 1990s the UK claimed to be leading on the grounds of numbers of companies. When the UK was overtaken in 1999 by Germany on company count, UK apologists moved the argument to one addressing the total industry size. But on the measure of real company value (i.e. stock market capitalisation, which as we have discussed is the central driver of the industry), Figures 2.8 and 2.9 show that UK is a typical European country.

Comparing the UK or Scandinavia with the US is scarcely appropriate, given the size differences between them. We can compensate for differences in country size by comparing biotechnology company value per person, rather than per country, which is done in Figure 2.9.

Figure 2.9 addresses this by showing the aggregate capitalisation of all public biotechnology companies (including Amgen and Genentech) against the population size of several countries or regions. ('Europe' here is the common market (12), excluding Denmark because Denmark is economically and socially a Scandinavian country). Europe does substantially worse than the trend: Canada, Australia and Scandinavia all 'punch their weight' for their population size compared with the US. The argument can be made that Canada benefits from proximity to the US capital markets and a common language, but this can hardly be said of Scandinavia.

3
Europe Falls Behind

The European biotechnology industry is, therefore, smaller, less productive and growing far more slowly than its US counterpart, and this malaise is not related to Europe not being part of the US, nor is the UK's relatively poor showing because it is smaller than the US. The companies are also 'younger'. VC-backed biotech companies typically start out to do research, discovering a product and only when that product is mature enough for a market do they fill out the organisational structure to include marketing, sales, manufacturing and the other functions expected of a fully fledged business. By the end of 2006, US public biotech companies had four times the employees of European ones, but only 1.3 times as many employed in R&D, that is, the European industry was far more focused on the early stages of the biotech business model [33].

So it is true that the European industry today is in some ways like the US industry 20 years ago. But the cause of this is not that it started 20 years later.

The 'biotechnology company' phenomenon started in California, with the creation of Cetus, Chiron and Genentech, and in Boston with the creation of Genzyme. Since those heady early days, the industry has grown to a multitrillion-dollar industry sector lead by major, profitable companies. Figure 2.2 provides a summary of a commonly used measure of 'industry growth' – the number of companies in the field, and illustrates the 'European lag'. Even though the number of European companies has 'caught up' with the US in the last decade, as we discussed above its capitalisation is the same as the US industry in the 1980s, and less than one of today's US giants. The argument can be made, therefore, that the European industry is 15 to 20 years behind the US, and it has taken a generation to transfer the idea of 'a biotechnology industry' to Europe (as it appears to have taken a further generation to transfer the idea from Europe to Japan, despite dire warnings in the 1980s from the US that Japan was about to overtake them in the Biotech field). This has been attributed to a lack of specialist VC in Europe [15, 18], lack of entrepreneurial spirit and other factors.

However, plotting the data in this way obscures the fact that at the start of the biotech decades, Europe and the US were very similar. High-tech company creation was happening in the UK as well as the US by the mid-1970s. Oxford Instruments, a company making high-specification magnets, was founded in 1959 and floated in 1983 on the back of the emerging field of magnetic resonance imaging (MRI) [34]. The modern form of the 'biotechnology company' – a company backed by investment that uses new science to research and then develop new products (usually in healthcare) – arose in Europe no more than 3–4 years after the US. The fraction of biotech companies formed per year since 1980 is similar in UK and US [24]. Prior to 1982 there was a 'jump' in company creation in the US, but Europe largely caught up in terms of rate in 1982–3 (Figure 3.1). Of the three companies often cited as

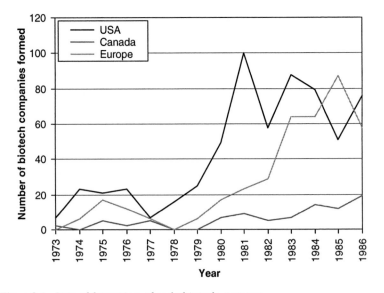

Figure 3.1   Rate of formation of early biotech start-ups
Note: Numbers of listed companies for US and Europe (UK, France, Germany, Benelux, Austria, Italy, Spain and Scandinavia), corrected for sampling error relative to Figure 2.2 by reference to Ernst and Young reports for 1993–5.
Source: Data from [26, 27, 29, 30] and from Bioscan Database 1995.

being in the vanguard of the industry in the early 1980s (Genentech, Cetus and Biogen), two were US, one European (see e.g. [35–8]). By 2000 the top ten were Amgen, Genentech, Serono, Biogen-Idec, UCB-Celltech, Genzyme, Medimmune, Chiron, Millennium [39]: only one of these is European, and that is a conventional pharmaceutical company (UCB) that gets into the list because .it acquired Celltech, which would otherwise not feature in the top 50.

(It is often forgotten today that Biogen, a paradigm of US academic spin-out made good, started life as a European company in 1980 [40], shifting to a transatlantic model through acquisition of a US research division in 1982 [36], and then moving its corporate headquarter to the US in 1983 [41], finally closing down the European rump in the late 1980s as it became obvious that Europe was not the place for a global biotechnology company. Biogen originally went to London investors for funding. Unable to get investment for a new biotech company in the UK, they eventually got Swiss backing and started operations in Geneva. It is ironic that the 'Gnomes of Zurich' as the newspapers called them at the time, famously conservative bankers from what is usually regarded as a conservative country, should chose to back Biogen when the powerhouse of entrepreneurial banking in London regarded it as too risky.)

This suggests a lag of no more than 3–4 years between US and European biotechnology industry at its inception. This is not at all surprising. Molecular biological research is, and was, global. Scientists moved readily across the Atlantic for temporary or permanent jobs, as we shall see later. Scientific papers were circulated round the world in days by post, within weeks through the conference network. Through ARPANET the scientific community was one of the earliest civilian adopters of what was to grow into the Internet as a global communication tool.[1] It was inevitable that, as soon as an enterprising scientist in California was successful in setting up a company to apply their science, at least some European scientists should seek to do the same. The current rather dismal picture of European biotechnology is not a result of us having started later. It is a result of what happened after the early 1980s.

However, even if biotech entrepreneurs were active and willing in Europe, unless their VC counterparts were also there ready to invest then a Genentech-modelled industry could not arise. It is widely believed that this precondition for the rise of a European industry was not met. Thakor is typical when he says [42],

Many have claimed that venture capital has been a significant factor in the rapid growth of new high-tech firms in the US, and that the relative absence of venture capital (VC) in Europe has been a note-worthy explanatory variable in understanding the relative absence of prominent high-tech firms in Europe.

Thus VC (or investment groups saying that they were VC) only started in the UK in the 1990s. It had been established in the US since the 1960s. With such a tight link between VC and biotech, it is inevitable that the biotech industry will be younger as well.

This is fallacious for three reasons. Firstly, it assumes that 'venture cap-ital' has only been available in the UK since the 1990s. In fact, there was early stage development capital in the UK before the 1980s, and it did invest in biotechnology. Indeed, investment in early stage companies, and specifically technology companies, is a long tradition in Europe. 3i was founded (as Investors In Industry) in 1945 for private investment in new manufacturing concerns, Technical Development Capital was put-ting £1 million/year in new technology projects from 1972 [43], and by the early 1980s there were enough companies calling themselves 'ven-ture capital' to found the European Venture Capital Association in 1983, which is funded by fees from VCs and similar groups [44]. The idea that investment in the early stages of technology-based companies in the UK only started with the rise of major VC houses in the 1990s is a revision of history by those houses.

VC investment (or rather, investment that styled itself as US-style VC) was available in the UK from the start of the biotechnology era [21]. Biotech Investments Limited (BIL) started investing from London in the early 1980s, with a $60 million fund (equivalent to a $150 million fund today, a good size for a fund investing in early companies which, of necessity, all companies were at that time). However, all their first investments were in US companies, except Celltech [18], in which they invested $12 million in 1981 and 1982. BIL's average investment was $3 million, equivalent to $6 million today, a decent investment level in early stage companies of this sort. By 1982 BIL had invested in 12 start-ups, and by 1984 BIL, Technical Development Capital, Prutech, Cogent and Advent were all investing in biotech out of London. Although the totality of VC in the US was much greater, the number investing in biotechnology companies was comparable on a *per capita* basis to this investor base in London [43]. In 1981 Technical Development Capital invested in start-up biotech Cambridge Life Sciences [45] among others, investing £1 million (equivalent to about £2.5 million today).

The amounts invested were also similar. The first companies were founded with similar initial ('Series A') investment to their US counterparts.

Celltech received £10 million initial funding [46], equivalent to a £22 million Series A today. British Biotechnology was set up in 1986 with £2.6 million, and a follow on of £8 million, equivalent to £4.7 million and £14 million respectively. Thus, the drive to create new businesses in Europe was roughly contemporaneous with that in the US, and *initially* they were set up on roughly equal scientific and financial terms. The US and the European 'biotech' industries entered the 1980s on level pegging.

However, it is true that the early biotech entrepreneurs in the UK saw the US model of VC-backed start-ups, and realised that it could not be replicated in Europe. A far higher proportion of biotech start-ups in UK relied on the entrepreneur's own cash than in the US [47]: this is true in the 2000s as well as the 1980s [48]. For growth capital, the majority of entrepreneurs in the 1980s turned to corporate ventures. Indeed, there has been a long tradition of institutional speculation in high-tech start-ups in the UK, some of which was used in the early biotech growth in this country. Institutional investors could then have chosen to provide development finance for those companies – that most of them did not chose to do so is a reflection of them.

This brings us to the second fallacy – that the only source of start-up capital for biotechnology companies in the early days of the industry was VC. In fact, VC was not the original source of start-up capital in Europe. Grand Metropolitan (a hotel chain), Air Products, Ciba-Geigy, Eli Lilly, ICI, Schering Plough, Monsanto and British and Commonwealth Shipping had all funded biotech start-ups in the late 1970s and early 1980s [35, 49], including investing in Biogen and Celltech. The British Technology Group (BTG) was active at this time investing in the exploitation of new discoveries from the biotechnology disciplines, a hybrid between 'seed funding' and technology transfer in today's terminology. By 1982 BTG had invested in over 40 UK biotech ventures [50]. We shall return to this point in a moment.

This early industrial support was not a uniquely British phenomenon. Cetus, founded in 1970 and well ahead of the VC-driven biotech boom in the US, was originally backed by Standard Oil [51]. Advent Ventures, one of the early investors in UK and US biotechnology, was created by Monsanto to invest in what they saw as a growing area. Other, mainstream VCs only got involved years later when the model had been established. Thus, in the UK and the US, VCs were actually following a trend established by industrial investment, not creating an industry.

This is a common theme. Mason [52] examined whether VCs created high-tech clusters or followed them in Canada, by looking at whether VC activity preceded or followed company creation: his data (Table 3.1)

Table 3.1   VC and technology company formation activity in Canada

Dates	Technology product firms	Technology service firms	VCs in area
Before 1988	19	3	1
1988–1996	20	11	2

Source: Reproduced from [52] with kind permission of Elsevier. Copyright Elsevier (2002).

show that VCs follow industrial trends, they do not create them, despite claims to the contrary.

Avnimelech [53] studying the rise of Israel's VC industry in the late 1980s to 2000, found that the precondition for the VC firms being founded was the existence of a large pool of entrepreneurial scientists and engineers, a rising number of new start-ups, individuals coming back to Israel from the US wanting to start companies and new high tech business models being tested in real life. In other words, the industry comes first and VC investment follows. This is not unique to biotech. VentureSource and Ernst and Young reported that VC investment in 'clean technology' companies doubled in 2006 compared to 2005, some years after the political and industrial growth of the sector had been under way, and about the time that 'An Inconvenient Truth' demonstrated that global warming had become an established political fact by winning an Oscar [54].

Thirdly, European companies could go to the US for VC funding, and indeed many did and continue to do so, although the barriers are significant. Biogen went to the US for investment, and found the environment there so much more congenial that the whole company migrated there over a period of five years, becoming a US company. It is now cited as one of the most successful start-ups in the Boston area. Others either sought capital in the US but stayed in the UK (like Cantab, Ethical Pharmaceuticals and Xenova [55]), or like Biogen moved to the US completely (like Scotgen, Amylin [55], Pharmaceutical Proteins [56], IDM [57] and others) This is not solely a trend of the 1990s. In 2006 Cyclacel completed a reverse takeover of Xcyte, stating that the reason that they did this rather than seek funding in the UK was that suitable funding was just not available in Europe [58] (I discuss this case further below, as this is not the whole story). Biovex, IDM, Micromet, Nordic Bone also followed this route, despite the barriers of cost, distance and the greater regulatory demands imposed by Sarbanes-Oxley Act in the US [59].

But lastly, even if it were true that companies could only get funds from VC and that VC was not available in 1980 to support start-ups in

Table 3.2 VC capacity versus population

Region	Population (millions, 2005, rounded)	VC firms	VC/1m pop	VC funds ($ million)	VC funds/ 1m pop
UK	57.5	110	1.91	4500	78.3
USA	248.7	550	2.21	20000	80.4

Source: Data for late 1980s to mid-1990s, from [47, 60]; Population data from CIA World factbook.

Europe, and moving to the US was impractical, it is only true in the very early days of biotech.

It is not the case that, in the UK at least, that there is less VC overall compared to population size (Table 3.2). To say that 'UK VC is young because it started in 1980' just highlights that it started at the same time as the biotech revolution. As noted, BIL raising a total of $71 million for biotech investment in 1981, enough to kick off a dozen biotech companies on its own. There was a boom in VC fund creation in the UK from 1985 [61]: independent UK VC funds raised £7 million 1979, £166 million in 1984, £222 million 1985. By the late 1980s, both the aggregate amount of VC and the number of firms is comparable on a per-population basis to that in the US [47]. There was a relative lack of VC in France, but then there was a relative excess in Canada (compared to the US on a per capita basis): one does not, however, hear the US complaining that Canada is a better place to start a biotech company because there is more VC there.

However, this VC funding was not put into high-tech start-ups of any sort. Most UK VC has traditionally gone into 'conventional' business – newsagent chains, car valeting services, take-out food companies – rather than into high technology. Gompers [62] notes that 'VC' has quite a different definition in Europe, which includes MBO, mezzanine and late stage financing. In US, it means early stage risk capital. Funds were raised under the 'VC' banner in the UK, but were not used in that way.

From an investment perspective, none of this would matter if the biotechnology industry made money for its stakeholders. This is distinct from the argument about size – tiny industries can make money for a small number of people. In Europe this has not happened, (with one exception I will discuss in Chapter 11), again unlike the US industry, where Genetic Engineering News (GEN) regularly published lists of 'molecular millionaires', individuals who, as founders and CEOs of biotechnology companies, have made substantial amounts of money. GEN's list usually excludes bankers, venture capitalists and financiers who have made money from biotech, recognising that these people do not create companies, no matter what they say.

It is sometimes argued that discussions of money made miss the point. Academics do not usually become involved in founding or supporting biotechnology companies purely for the money – they want to see their science developed, their reputations enhanced and people helped. But creating and running a business takes years of commitment. If the academic is going to make more money out of writing a book than working on their project, their altruism will be opposed by the need to pay the food bills. And others working with them – early investors, managers, technology transfer offices – will be dissuaded completely from working on the company (see [63–4]).

The biotech industry is meant to be a business – it makes money from its transactions for its stakeholders. If it does not do so, then it must be counted a commercial failure.

Maybe the biotech industry 'is not making profits yet', with the implication that it will do when its science or products mature in 5, 10 or 20 years' time. I have no personal doubt that by 2050 'the biotech industry' will be a huge, profitable and potent part of the economy. However, this argument misses the point. People who *invest* in biotech companies do not want to have a return in 40 years time. Managers and entrepreneurs do not want a return after they are dead. For them the industry must make money in a commercially plausible timescale, and it is reasonable to say that since the creation of Biogen and Celltech in the 1980s the European industry has had time to make money for its investors and entrepreneurs. However, it has not done so.

The failure of the European industry to make money for its investors is hard to track. The rise and fall of the public companies is well documented, but tends to follow the fate of an industry norm. The profit (or loss) of public investment vehicles is dominated by the overall movement in biotech stocks, driven by many factors from the marginally relevant (such as a notable failure in one company's clinical trial causing the whole sector to drop) to the almost completely irrelevant (e.g. the 7 July 2005 bomb incidents in London caused shares in Vernalis to drop 3%, reflecting a market move away from risk, but shares in Acambis, a company that makes vaccines for the US for 'homeland defence', to rise 11%).

Private companies should be more revealing. Here the synergy or symbiosis between venture capitalist and entrepreneur should be able to develop the value in a company away from the glare of public attention and the demands for quarterly returns, taking some great science and turning it into a saleable equity asset. However, these are private

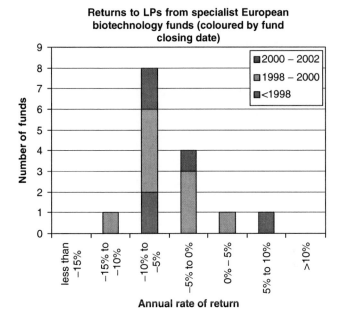

Figure 3.2 VC returns from European biotech VC funds
Note: Returns on specialist European VC funds – note that the funds often invest in some US companies. Funds are categorised into those raised in 2000–2, 1998–2000 and before 1998. Valuation dates were in 2005 and 2006.
Source: Data from Limited Partner annual returns in 2005 and 2006.

transactions, and few European venture funds that invest in Biotech admit to the profit or loss they make for their investors. VCs typically over-claim their returns: actual returns are almost always below 20%, compared to 30%–50% claimed [65–7]. Tyebjee comments that 'A remarkable 74% of the BVCA survey respondents agreed with the statement that "UK venture capitalists had oversold the potential returns of their industry to institutional funders"' [67]. But even if they only performed as well as the stock market, that would be a useful yield, as the public stock market does rise historically (albeit no faster than many other forms of asset, as shown in Figure 3.4).

The reality for the value of VC funds in European biotechnology is astonishingly different from the spin. Some limited data on the performance of funds is available from public sources, and suggests that the funds do substantially worse than the historical average for the stock market (Figure 3.2).

The average return is −6%, that is, a *loss* of 6% per annum. On a ten-year fund, this equates to loss of 50% of the invested capital. VC not merely does worse than putting the money in the bank, it does worse than storing it in bank notes under the investors' desk. The period has been chosen specifically to avoid funds that had to realise in the 2000–4 period, which cover a highly abnormal market for sale of any high-tech company shares on public markets, and covers funds raised throughout 1997 to 2002, and hence investing before, during and after the stock 'bubble' of 2000. This should therefore represent VC fund performance across a typical cycle of the market.

This result is at such divergence from the stated view by the industry that it typically makes 15%–20% annual returns on investment, and that 40% returns on early stage VC are not implausible, that I should take a few lines to justify that Figure 3.2 is an accurate reflection of the reality of European biotech VC returns by dealing with a couple of obvious objections.

The first objection (and no doubt one which will be widely expressed) is that this is just an opinion, not the objective reality. This does not stand up, as this is not my analysis: it is the summary of the audited accounts of a number of public companies which have invested in venture funds. Whether independently audited accounts reflect reality is a matter that VCs need to take up with the relevant accountancy professional organisations, and one we will touch on later.

But why should returns audited in this way be so low, when VC claims are for much higher (or, at least, positive) profits? Anticipating what is to come, it is in part because of how VCs report their funds' activities. Funds of a particular size do not get all their investees' money at the start: They draw it down (i.e. get the investees' to actually hand over the cash) when they need it. If the return is calculated on a fund as the return on the amount invested *when* it is invested versus the amount gained back from realisations when the fund sells the shares it has bought, or that same amount at the end of the fund's life, then you get the Gross IRR (Internal Rates of Return). However, this is not what the investors get, because the VC management group requires management fees throughout the fund's life (as summarised in cartoon form in Figure 2.1).

The result of this is illustrated in Figure 3.3. Here £40 million fund is invested in just four companies: £10 million in each in the first, second, third and fourth years of the fund's life. One of these yields an exit that returns £50 million to the investor, the second a return of £75 million in the last year of the fund, the other two investments are written off. So £40 million invested, £125 million realised over ten years, an apparent return of 367%, or 12% per annum. If the Limited Partners calculate

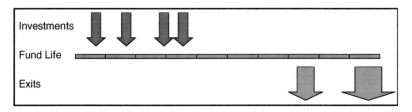

Figure 3.3   Simplified fund life cycle example

the IRR based on when they actually handed over the money, the IRR is 15%. If the fund pays its returns to investors when realisations occur (rather than at the end of the fund's life, something that was being required increasingly by US investors in the late 1990s [12]) then the return looks like 18.5% But after the fund fees are paid (which are paid from day 1) and profit share is accounted for, the return is 8.5% under half the most optimistic of those figures. VC funds will of course emphasise the 18.5% figure or '£40 million in, £125 million out'. The reality for their investors is, however, less rosy. If the second exit was at cost to the VC (i.e. realised £10 million *for this investor*: if this investor is only one among several this might still represent an IPO valued at £50 million, quite common in Europe), then the IRR on a cash-in/cash out basis is 9%, but the actual return is −1%: the fund makes a loss. (This is rather better than historical averages suggest, but we will assume our VC is a high performing fund manager.)

So 'raw' figures can provide a very positive picture, but the picture the investors in VC funds actually see on their balance sheet is much less attractive. It is the latter which is reflected in Figure 3.2.

Secondly, these funds are mostly quite young: it could be argued that funds often appear to be 'under water' in their early years: as investee companies develop, their value increases dramatically and the overall value of the fund increases. So, the argument might go, these calculations undervalue the investee companies' inherent value, while including the full cost of running the fund, and hence substantially undervalue the fund as a whole. But this is not how these valuations are achieved. Most Limited Partners (LPs) explicitly state that their valuations are done to the EVCA/AFIC/BVCA[2] valuation guidelines [68]. The guidelines are aimed at allowing funds investing in private companies to report a 'fair value' to their investors, and the guidelines specifically state that good accounting principles are included as part of the guidelines. The guidelines include provisions for estimating value if the fund is a minority share-holder and unable to influence the decision to exit, but as we shall see that is not the case for VC investment: they have effectively complete control over exit decisions (although such control can be split between several, conflicting VCs). In the case of recent investments, the guidelines state

> Where the Investment being valued was itself made recently, its cost will generally provide an indication of Fair Value. Where there has been any recent Investment in the Investee Company, the price of that investment will provide the basis of the valuation. . . . [For] early stage enterprises or enterprises without or with insignificant revenues, and without either profits or positive cash flows . . . the most appropriate approach is a methodology that is based on market data, that being the Price of a Recent Investment. This methodology is likely to be appropriate for a limited period after the date of the relevant transaction. The length of the period for which it would remain appropriate to use this methodology for a particular Investment will depend on the specific circumstances of the case, but a period of one year is often applied in practice. After the appropriate limited period, the Valuer should consider either whether the circumstances of the Investment have changed, such that one of the other methodologies would be more appropriate, whether there is any evidence of deterioration or strong defensible evidence for an increase in value. In the absence of these indications the Valuer will typically revert to the value reported at the previous reporting date.
>
> [68]

In other words, investments are carried at cost until a further investment re-values the company as a whole, or until something very notable happens. Closed rounds, small investments and bridging investments are not considered useful for this, so revaluation only occurs when a major round with new investors is achieved, which is usually two years or so after the original VC investment, that is, 3–4 years after the fund that was invested has closed. The method provides for discount of the cost of investment if progress after investment has been below that expected when the investment was made (milestones missed, further financing not raised etc).

Thus the standard approach is to assume that the value of a private investment is the cost of that investment unless something bad happens. As a default, then, VC funds should have a growth of 0%. Negative growth, which is what is found in the majority of cases, must represent failure of investee companies to meet expectations within the timescale being examined, not failure to reach their final goal within the ten-year fund investment strategy.

This, however, is not a failure to meet revenue targets, but a failure to maintain their value: the caveat excluding 'at cost' valuations for companies with significant revenues does not apply, because biotechnology companies rarely have revenues, and indeed VCs try to prevent them from having revenues as we shall see. As I noted above, determining the fair value of an early stage biotechnology company is very difficult. It requires deep knowledge of the progress in the underlying research and development, understanding of the strength of the patent portfolio, detailed knowledge of their specific business dealing and the general business environment, and an ability to predict how the management team will develop all those features in the future. Clearly a standard audit by a general accountancy company will be incapable of evaluating that. In fact, the only entities able to judge a company's value without the detailed and exhaustive research performed during due diligence studies (which often turn out not to be all the exhaustive or diligent [69]) are the company and its VC investors. Clearly, neither is inclined to be pessimistic, so a fall in value means that an optimistic observer, with every interest in maintaining the appearance of value of the company, has had to admit that it is less valuable that it was. (I will discuss the implications of this for the VC's own business model in Chapter 11.)

Thirdly, it could be argued that the 1995–2005 period was a poor time for investments as a whole. This is true of some periods within it, but not sufficient to explain the results. Figure 3.4 shows what returns could be expected if the money invested in the VC funds analysed in Figure 3.2 had instead been invested at the same time in a NASDAQ tracker fund, and audited at the same time as the VC LPs.

The conclusion from Figure 3.4 is that putting your money into a NASDAQ tracker over the same period would have done no worse, and possibly better, than investing in VC. (More specifically, Figure 3.4 shows that if you bought NASDAQ stock in April 2000 then you would have sold at a loss any time afterwards, but otherwise you would have made a profit: VC investments show almost exactly the same pattern.) The average return for NASDAQ stocks, picked to match exactly the buy- and sell-profile of the VC funds, is 1% gain per annum – scarcely stellar, but better than a 6% loss. Interestingly, NASDAQ shows *more* volatility over this period than the aggregate VC funds, suggesting that the VC funds are being very conservative.

This is not a unique result. Booth [70] reports five-year returns across biotech VC 2000–5 (the tail end of one boom and the following bear market) to be generally dismal, and to rack up losses of 10.8% per annum compared to 2.2% falls in the NASDAQ over the same period.

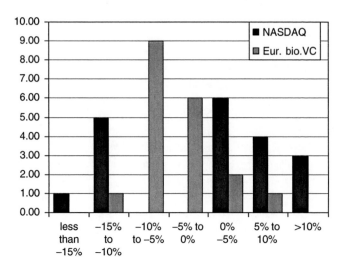

Figure 3.4   NASDAQ comparators for VC returns
Note: 'Funds' of NASDAQ tracker stocks bought in Q2 and sold in Q4 of the same year as the VC funds analysed in Figure 3.2, compared to Figure 3.2 data.
Source: NASDAQ returns over the same period as Figure 3.2 data.

Given this appalling record, why do people invest? We will come to that at the end, but we can note that returns of this sort are not typical of all VC, or even of the majority of biotechnology VC in the US. VC groups made excellent returns in 1960s and 70s. By 1980s returns were declining [71], but were still positive. Gompers [62] claims returns in the range 20%–30% for VC funds in the 1990s – these, however, were VCs own claims of returns, that is, were 'raw' returns before subtracting operating costs and profit share. Large European VC made decent profits in the period considered in Figure 3.2: 3i reports 17% profit on its venture operations in 2005 [72], GIMV report that by 2005 profitability, which had fallen to a 4% loss in 2001–2, was back to 15% [73]. So European biotech's performance was not the result of underperformance of VC as a whole, or of investment markets as a whole. Nor was it a result of the underperformance of the biotech sector. With the exception of the period 2001–3, and November 2007 onwards (which is after the periods considered here) biotech stocks have performed solidly since 1995 as an average. No, this was specifically that European biotech VCs have lost money, in some cases spectacular amounts of money.

This is not to say that the VC groups themselves have not made money. As I will analyse in Chapter 11, venture management teams have made significant money from their work. But the ultimate investors, the groups that give the management teams funds to invest, on average have not.

Financial 'failure' for a company founder, entrepreneur or senior manager is a more complex issue. Many entrepreneurs believe they have failed when they do not achieve their personal goals, which, in a financial setting, are often vaguely stated to 'be rich'. (Although in an industry where a 28-year-old postdoctoral researcher with a first class undergraduate degree, a PhD, and four years' laboratory work experience at a world-class research institution can be expected to be paid around £35,000 per annum, their idea of 'rich' is not necessarily very ambitious.) An analysis of entrepreneurial goals is also complicated by the other goals that entrepreneurs have. Few people start a biotechnology company purely to make money – money is more easily made in banking, law, for the technically adventurous in software, communications, the cutting edge of silicon fabrication, for the commercial in sales-driven businesses. Biotechnology companies are usually started by people with a vision that combines science and commerce, and personally therefore also combines scientific, personal career and financial goals, as well as a desire to work with friends, be independent and so on. The motivation of entrepreneurs has been studied in a variety of other contexts and is found to be equally mixed. Failure of biotech entrepreneurs to be able to articulate their goals in numerical, financial terms is not a 'failing' of the biotechnology field, although it would no doubt help the entrepreneurs if they could be more precise about why they pursued the career choices they did.

So asking an entrepreneur whether they are a failure is not helpful. Rather, we should ask for objective business criteria of failure, and for the entrepreneur or company manager the obvious criterion is failure to make money.

The published stories of entrepreneurial success in biotechnology, as in any other field of industry, are biased heavily towards those that *have* made significant money. Most stories about this are anecdotes retelling the high profile successes of founder academics who make substantial sums through sale of equity in the companies that they founded. But this is totally atypical. The UK 'serious' press has regular articles describing millionaire professors or business boffins: one such article by Jessica Shepherd listed about 34 life science Professors who had made substantial sums from life science entrepreneurship, although four of these had not actually made significant money, and this spanned a 30-year period (it included Ken Murray who was alleged to have profited from the creation of Biogen in 1980), that is, about one a year [74]. An earlier article [75] listed 20, of which only 13 had actually made the cash they were alleged to have made, and eight of these were in common with the list compiled by Shepherd. So the population of genuinely rich academic spin-out founders is very small, and not representative of all start-ups, let alone all academics. My informal request of UK investors and entrepreneurs to name any European biotech scientist millionaires came up with a handful of names, one of which (Charles Cantor, co-founder of Sequenom) is actually American.

In addition, the share structures of many companies are made so opaque by multiple rounds of investment on different terms with different share classes, preference structures, anti-dilution clauses etc. that calculating the value of a company founder's shares at company IPO is extremely complex: this is an important issue we will return to. A more valid approach is to conduct a prospective trial, asking, of a suitable sample of entrepreneurs today, how many make significant return for their efforts.

As academic scientists in Europe and the US are being encouraged to commercialise their discoveries and expertise, by forming spin-out companies or by licensing technology, and being encouraged to do so in part at least because of the financial rewards that are alleged to accrue, I did a fairly detailed study of how much such scientists could be expected to make from their efforts, compared to how much they would make from other exploitation routes [22]. This is separate from whether investors make money, or whether the creation of spin-out companies is worthwhile for the national economy as a whole.

The study calculated the likely overall financial reward, summed over their whole career, for a University scientist with a good idea or discovery that they thought might have potential commercial value, to be gained from

(i) leaving academia,
(ii) commercialising specific inventions though intellectual property (IP) licensing,
(iii) formation of a spin-out company (specifically, a spin-out funded by VC),
(iv) commercialising the academic's expertise and know-how through consultancy and
(v) writing for markets that pay the author (all scientists write technical papers, but these almost never pay their authors – indeed increasingly they require the authors to pay the publishers).

The outcome of the model is shown in Figure 3.5.

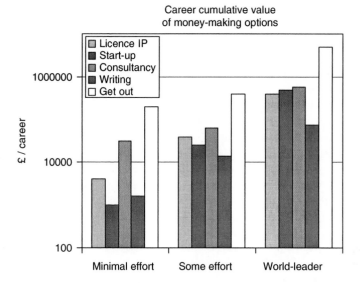

Figure 3.5   Scientist career total earnings from different approaches to science commercialisation

Note: Y axis – total likely earnings from commercialisation of your ideas if you are a UK academic scientist. Earnings are categorised according to whether you chose to licence intellectual property through your university to an existing (usually large) company, leave academia, form spin-out company(s), consult or write a book about it.

Source: Reproduced from [22] with kind permission of Palgrave Macmillan.

The analysis looked at three levels of effort the scientist might chose to take. 'Minimal effort' means that they go along with what the university, publisher, venture capitalist or consultee company wants, but does not pursue business aggressively. 'Some effort' means that they pursue commercial outcomes of their research, and put significant time into them. 'World class' means that they do this, and are also in the top 10% worldwide in their subject.

The best option is clearly to leave basic research and to join industry, climbing the ranks of industrial management. Assuming a creative scientist does not want to do that (and most do not, at least at the start of their careers), in all cases it is more sense for a UK academic to sell their ideas as a consulting business than to start a VC-funded company. For the majority who do not aggressively pursue a career in commercialisation it is true that if they have a good idea they will make more writing a book about it than starting a company,[3] and if they do start a company the average amount that they can expect to make around £25,000 from their effort, ideas, expertise and up to five years' extra-curricula work.

Figure 3.5 does not take into account the effort involved in the activities, or the opportunity cost of each. Informally, I observe from personal experience that founding a company takes 3–10 times as much work time as writing a book,[4] which suggests that of these two options the latter is a better investment under all circumstances. Both are extremely poor rewards on a per hour basis compared to consultancy. The additional advantage of writing is that you get the ego-gratification of seeing your name on bookshop shelves, whereas your name can be rapidly expunged from the corporate history of the company you create.

Surveys from the US (e.g. [76–7]) suggest that earnings from spin-outs among US academics are on average substantially more than in the UK if the academic retains any equity stake at IPO (which is only true for 40% of IPOs [76]). However, my informal discussions with academics in the US suggest that the amount they can earn from start-ups still does not reach that of consultancy.

Company executives fare better if only because they are paid, for as long as they are employed, however, this rarely adds up to more than £350,000 (three years (see 7.9) at £120,000 for the CEO). Thus the claim by Merlin Biosciences (or more accurately, Merlin's chairman) [78] to have created 50 biotech millionaires through investing in biotech companies seems unlikely unless the investors and fund managers are included in the count: very few people become millionaires from being associated with a VC-funded biotechnology company in the UK.

So, the European biotechnology in recent times (i.e. 1990 onwards) has failed to grow despite a promising start, has failed to make its ultimate investors any money, and in most cases has failed to make the entrepreneurs and managers responsible for creating, building and operating companies any money. This is in complete contrast to the claims of many government apologists and investor groups in the industry. These failures are linked, of course – a company that fails to grow and develop products will not make anyone rich.

Why did this happen? A wide range of reasons are suggested (often, oddly, by the same people who claim that European biotechnology has been an outstanding success), and there is no one, unique factor. For example, the domestic economy in the UK must take some of the blame. Investing in domestic property in Cambridge, UK (i.e. buying an existing house and sitting in it), has routinely brought similar returns to investing in high growth stocks such as NASDAQ tracker for almost every year in the last half century except 1990–3 (when the UK housing market slumped) and 1998–2000 (when the dot.com boom boomed), and is not subject to Capital Gains Tax. This drains capital from the productive creation of companies into the economically useless activity of passively existing. But, these are relatively minor factors (this one does not explain why biotech companies did not spring up in Manchester to rival the Bay Area, for example, nor why the Bay Area's high house price inflation has not discouraged entrepreneurship there).

My contention in this book is that the behaviour of investors is a major reason for this industry's underperformance. This is not only that investors do not invest. It is how they invest, what they invest and what they do afterwards.

As this is a not an undisputed view, I will dispose of some of the other reasons commonly given for the underperformance of the European biotechnology industry in the next chapter, before analysing how the behaviour of the VC industry in Europe has caused this industrial dwarfism.

4
Why Does Europe Do So Poorly? Some Inadequate Explanations

Thus the flower of biotechnology is a small weed in Europe compared to the giant redwood of the US. Why should this happen? Before delving into what I think is the main reason – the business practices of the VC groups that control the industry – it is worthwhile discussing the 'reasons' that are usually given, and showing why they are at most a minor cause of the relative lack of growth and frequent commercial and financial failure of the European industry. In this section I will explicitly leave out financial reasons such as the structure of the capital markets – we will deal with this in Chapter 6.

The most obvious argument, and one that is probably most common, is that Europeans are just not as entrepreneurial as Americans. This is sometimes said bluntly, sometimes dressed up in terms of 'cultural values'. It is true that on average Europeans do not have quite the materialistic drive that some Americans appear to revel in – but then neither do a lot of Americans. The issue is not whether everyone in the UK is driven to build a global biotechnology company, but whether there are 10,000 or so people driven to do so, and (more importantly) competent to do so. The answer to this is 'yes and no' – the reason for the 'no' I will address at the end of this book in Chapter 13, as it is a result of the industry's malaise, not a cause of it. The 'yes', however, is illustrated by a look at the industry of the 1980s. If VC was rare in UK, then of course the number of 'biotechnology companies', meaning companies supported by investment to generate new products from new science, will be small. If we extend the definition of 'biotechnology company' to include companies whose business is actually selling products based on the life sciences, then a very different pattern emerges. By 1988 there were ~450 companies that called themselves 'biotechnology companies' in the UK.[1] Most were performing conventional trading operations (manufacture

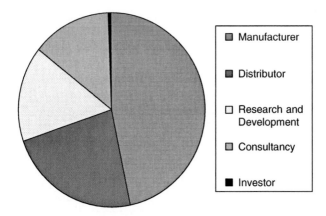

Figure 4.1   Business models for biotech companies in 1988
Note: Manufacturer – company that primarily manufactures the products it sells. Distributor – company that primarily distributes other's products. Neither category includes any significant R&D. Research and Development – any product company with significant R&D effort. Consultancy – any fee for service offering, including design, consultancy and contract R&D. Investor – investment or investment management group.
Source: Data from [79].

and distribution of their own products or distribution and sales of products manufactured elsewhere – see Figure 4.1). Most of those performing R&D were also selling products to their customer base, and funding R&D out of this. So the number of investor-supported, R&D-based companies was very small.

This is a common theme in UK biotech – Freel [60] notes in the late 1990s that 56% of innovative UK firms were funded by entrepreneur's private savings rather than VC, and self-generated profits the single most common source of continued funding.

The environment was harsh – survival rates for biotech companies then, as now, were not good, and an average half-life of 3.5 years seems to be a constant of the industry (VC-funded and otherwise) since the 1980s (Figure 4.2).

This is a rather faster fall-off rate than found by Deakins [80] based on VAT deregistration statistics for the generality of small companies, suggesting that biotech start-ups are more volatile than average, even by the unstable standards of new SMEs. And yet some of these companies survived 15 or more years as profitable entities, proof if any were needed of the practical business acumen of those running them. Of the others, they kept coming back, proof of indomitable entrepreneurial spirit (or incurable optimism, itself a key entrepreneurial trait).

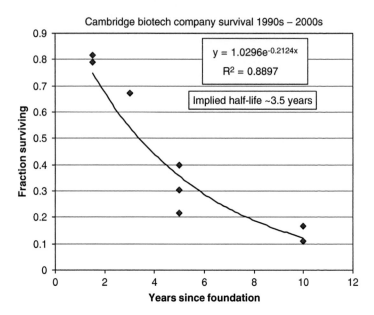

Figure 4.2   Survival of Cambridge area biotech companies
Note: Survival of companies over a variety of timescales from 1988 to 2003. Solid line is a curve matched to negative power formula.
Source: Data from directories, Cambridge Enterprise.

The spirit of entrepreneurship was not universal. Fishlock [50] tells of a city investment house that wrote to 1000 UK research institutions and researchers in electronics, soliciting proposals for new businesses. 'It received back a handful of replies, half of them abusive' [50]. He goes on to comment 'Little wonder, then, that most academics in the late 1970s and early 80s reacted with scepticism, if not outright horror, to corporate offers to fund biotechnology research in exchange for commercial rights. Today, observers on both sides of the divide concur that, while such joint research may not be for everyone, most of the early fears were unfounded.' This latter comment, however, was on the *US* environment, not that in the UK. Investors chose to see the pro-entrepreneurial academics in the US, the anti-entrepreneurial ones in the UK and judge the two countries on those preconceived criteria [81].

True entrepreneurship – the drive to work at the creation of new commercial ventures, was therefore alive and well in the UK in the life sciences in the early 1980s. The drive to create new companies from academic bases was much rarer, but as we shall discuss in a moment, no one in the

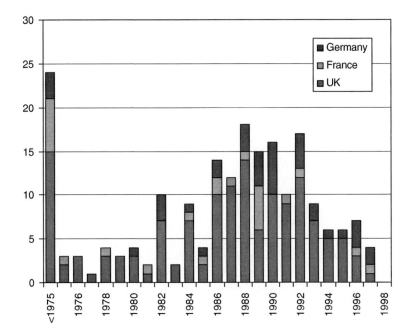

Figure 4.3 Date of foundation of 1998-vintage companies
Note: This represents 142 out of 269 UK companies – the other 125 did not list their date of formation.
Source: Age of companies recorded in BioIndustry directory 1998.

UK believed then (or believes now) that academics could make good business people anyway. It is entrepreneurship among true business people that is critical. This has flourished since – biotechnology industry directories (mostly UK-published, which therefore under-representing other countries) show steady flow of company creation since 1980, as illustrated in Figure 4.3, which shows the date of foundation of companies falling into the broader category of 'biotechnology' in 1998. As most companies active in 1988 would have vanished by this date, the record is quite impressive.

Even VC professionals had to concur that by 2000 there was a plentiful supply of entrepreneurs in the UK that they wanted to back [82]. The lack of 'entrepreneurship' is certainly a myth today, and was probably a myth in 1980 and before.

The second argument provided for the lack of substantial, VC-backed companies in Europe is that there are fewer companies worth investing in [47]. The herd-like surge to invest in biotech in 1999–2000 suggests that these scruples about investment quality can be overcome rather easily, but we must admit that there may be fewer 'good' companies in Europe than in the US. Many companies set up in the late 1990s had business models that had no hope of succeeding, or even surviving for long. Are the European companies just not up to the mark of US ones from the start?

This has been a common refrain from the early days of the industry. Biotechnology Investments Ltd (BIL) complained publicly of the poor quality of the UK companies they were shown in 1980–1, at the start of the biotechnology corporate era, explaining this was why all their 8 investments had been in America [50]. But they received 68 proposals from the US, 16 from the UK (out of a total of 98 – a tiny deal flow compared to the 1000 average a good VC would receive today) [50], exactly in proportion to the population, and the chance that none of the eight out of 98 they invested in was one of the 16 from the UK is 25%, not nearly enough to suggest other than pure chance that the ones that caught their eye were from across the Atlantic. They saw in their own figures a bias that was not there.

This argument has continued, though, and gained substantial force in the 1990s when a series of government initiatives stimulated the creation of new companies from academic research. Although there were over 1000 biotechnology companies in Europe in the late 1990s, most of them were not commercial propositions, and although formally constituted as companies were not, and could never be, businesses. The companies had no commercial goal other than to raise investment finance to do more research.

However, this should not result in a small UK or European industry over all, providing VCs are willing to invest appropriately in the few companies that *are* commercially viable. We have not defined the success or failure of European companies as the number that *have died*, any more than you define the population of mosquitoes in a Louisiana swamp as the number that have died. You count the number, size and aggressiveness of those that are there to bite you. Similarly, if these biotech companies are poor investment propositions, because their commercial rationale is flawed, then they will not get investment. Their existence should not suppress the numbers of companies that do have a strong commercial rationale and, for those whose rationale includes a need for investment, can provide a strong case for investment. The uncommercial, state-sponsored companies are ineffective competition for companies that do have a commercial future for either investment pounds or commercial market share. Of the many things that VCs complain about, the inability to read all the business plans that arrive on their desk is not one of them. Almost 95% of the applications for finance are rejected [21], because VCs can only fund a small number of deals. Typically, therefore they scan a large number of opportunities (the volume of which is called *deal flow*), rapidly evaluate most and reject them, and then concentrate on the small number that have potential [67]. Thus academic spin-out 'non-companies' will only diminish the health of the industry if VCs are so foolish as to believe they have a commercial future and invest in them: otherwise they will simply be a small added overhead on the (usually low cost) 'deal flow analysis' staff.[2] If VCs *are* so foolish, this can scarcely be seen as a failing of the *good* companies that exist, nor (realistically) of the academics who see a soft source of money in an era of ever-tightening academic budgets and exploit it to the full.

(The damaging effect of these companies if VCs do *not* invest in commercial successes will be examined in Chapter 11.)

Even if it were true that the *only* companies available for VC investment in the 1980s and 1990s in Europe were overly academic, there are examples to show that it is quite possible to set up an 'academic' company and make it a commercial success. DNAX was explicitly set up to mimic a University, with a follow-your-nose research policy and free publication rules, not driving to product development, but with a focus on excellence and on real world problems. This generated a good commercial result. DNAX was founded in 1980 with an investment of $4 million, and was acquired in 1982 for $29 million [83], an annualised rate of return of about 100%. If the business model worked then, it should work now *as a business* – the difference today is that investors would not *believe* that it could work, and so would not back it. It is not an *investment* business, merely a business discovering and developing products. Of course, DNAX was exceptional – most companies have more targeted research goals. But it illustrates that a company cannot be dismissed just because it is not carrying out product development yet.

Along with the 'poor companies' argument comes the 'lack of management' argument. European management is just not up to the job: the management in the US is more commercial, more experienced, more able to take some early science and turn it rapidly into a sustainable business.

This is very hard to refute. UK management in general is the worst in Europe [84], but that is in terms of running a stable business (and making life tolerable for their staff) rather than driving a company to rapid growth through investment and R&D. Mainland European managers are better than those in the UK, but different from US ones. US-centric

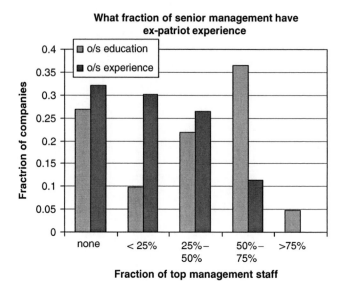

Figure 4.4   Employment mobility in European biotech companies
Note: X axis – fraction of senior staff in companies which have substantial ex-patriot experience (post-graduate degree education – a PhD or postdoctoral research for science, and MBA
for business – or at least a year at a senior job). Y axis – fraction of 93 European VC-backed
companies with staff with that experience profile.
Source: Data from company websites, gathered January 2007.

management theory often sees the US style as 'better', but there is no
evidence that this is so [85]. Who is to say which style or background is
best suited to creating a world-class biotechnology company?

Fortunately, we do not have to decide. The biotechnology industry is
global, like the science that supports it. The reality is that there is no
'European management' and 'US management'. Entrepreneurs tend to
act locally to wherever they start their career in entrepreneurship, as
why should they move? But managers routinely transfer, so gain skills
in either environment, and bring best home with them.

This is illustrated by the fraction of European biotech companies that
have management that has had experience outside the country in
which the company is based, shown in Figure 4.4.

Figure 4.4 summarises the mobility of the senior staff in those
European biotechnology companies that are VC backed (93 companies
in all). Around 70% of the top management positions (Chief Executive,
Scientific, Financial, Commercial and Operational officers) are filled with
staff with substantial education and/or commercial experience overseas.

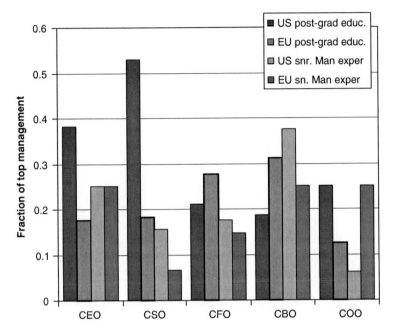

Figure 4.5 Expatriate experience of European CXOs
Note: X axis – senior officer position. Y axis – fraction with overseas experience, categorised by experience in another European country (EU) or in the USA (US), and by whether the experience is post-graduate education (post-grad) or commercial management (man).
Source: Company data set as for Figure 4.4.

Half the companies have expatriate experience in 50% or more of this senior band of staff.

The nature and geographic extent of the experience varies with the officer concerned.

As you would expect, CSOs more often have international education experience (postdoctoral research is usually considered as much education as employment, if only because of the very poor salary), CBOs have more international business experience (Figure 4.5). Only CSOs are more than 50% likely to have had overseas experience in any one category, supporting the very high mobility of scientists. But all of the top five jobs are likely to be occupied by people who are part of a global talent pool.

So biotech companies often recruit senior staff from outside their own country. Hendry [86], looking at European biotech companies, found that the majority of all recruitment (not just senior positions) was at the national level. However, the breadth of their research collaborations and contracts was international, again showing international presence and vision (Table 4.1).

So if European companies cannot attract good management, it is not because good management is not available locally, it is because they cannot attract good management from a global management pool. There is no such thing as 'European Management' to blame for the poor quality of the propositions. If a company cannot attract 'good' management it is because it is not an attractive enough proposition to attract management from this mobile, global pool.

We might note here that Asian countries might more reasonably claim to be limited by the supply of 'local' management – while many Europeans are happy to migrate to North America and visa versa, few would wish to make the very much more substantial cultural jump to Japan. And yet many European companies[3] have set up collaborations, offices and subsidiary management groups in the Far East, showing that the trade in management skills is truly global. The movement of Asian scientists to the US for work experience of a few years is so large that when the political changes after the September 11 bombings restricted Asian citizens moving to the US for temporary work there was serious concern that the result would cause the US to lose its pre-eminent position in biomedical research.

The more experienced (US) entrepreneurs have always taken this view: George Rathman, founder president of Amgen, the most successful biotech company of all time, commented:

> We've all known for some time that the management is probably not as significant as the science if you are going to be a leading biotechnology company. But an outstanding management team would not be there if they didn't have outstanding science.
>
> (Quoted in [87])

For an experienced manager, a 'good proposition' starts with top-notch science, but also includes investors who are willing to put the financial muscle behind the company to allow it to grow. Thus, the weak management argument is a *result* of the undercapitalisation of European biotechnology companies, not a cause of it.

There is another management argument, that founders, especially academic founders, do not have the experience to relinquish the reigns of power to experienced management. I will address the reason for this in Chapter 7, but I note that my discussions with US entrepreneurs and VCs shows that this is as much a problem in the US as in Europe.

Table 4.1 Where biotechs collaborate

| | Collaborate with | | | Research with | | |
	Other biotech companies	Pharmaceutical companies	Other companies	University	Government research institutions	Private research institutions
Local	0.5	0.17	0.13	2.11	0.79	0.466
National	1.55	1.1	1.5	3.27	1.82	1.166
International	1.92	2.17	2	1.2	0.34	0.533

Source: Data from [86], reproduced with kind permission from Elsevier. Copyright Elsevier (2000).

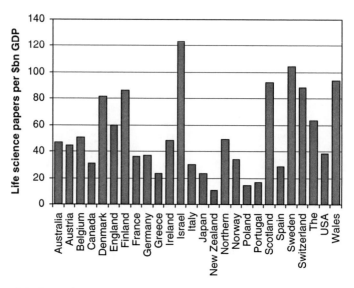

Figure 4.6   Scientific publications by country
Note: Scientific papers published in 124 specific life science topics related to biomedicine and limited aspects of veterinary medicine, 1995–2005 inclusive, as a fraction of GDP.
Source: Data collected by the author from Medline, 2007. GDP data from CIA world factbook.

Few dare be terribly explicit about the last potential non-financial category of excuse for Europe's poor showing, for fear of annoying political powers in government and in the Universities. But we must mention the suspicion that European life science just is not quite up to scratch. The US leads the world in Nobel prizes, despite the proud boasts of Europe's academic centres (Cambridge claims 19 Nobel Laureates, for example [88]). Surely this is a reflection of a level of excellence that must translate to commercial success.

This is surprisingly hard to measure. In terms of scientific papers published, Europe is on a par with the US. In the 1995–2005 period, Europe published around 375,000 papers in scientific journals dedicated to the biomedical sciences, to the US's 420,000 (in a period when the total output of scientific papers was slightly over 1 million). Compared to the GDP, the US record is average, and Israel comes out top (As shown in Figure 4.6, Israel is also a hot spot for entrepreneurial biotechnology, but not one I have extended this book to cover.).

However, publishing a scientific paper, no matter how grooming for the ego and helpful for academic career advancement, does not mean that a meaningful breakthrough in applied science has been made. This might be better indicated by spend on (necessarily expensive) biomedical research, and here the US does significantly better, as illustrated in Table 4.2.

There is a strong argument to be made that companies spun out from a research base which has received more investment will be stronger than those spun out of weak R&D bases. This, however, assumes that companies are spun out of the R&D base. Over half of biotechnology start-ups in the UK are not created this way, despite the government encouragement to do so, as illustrated by the example of the Cambridge area in the UK (Figure 4.7).

Table 4.2   Government spend on health R&D

Country	1975	1980	1985	1991
US	4199	4895	5611	7753
UK	150	128	305	398
France	266	329	412	408
Canada	117	114	147	203

Note: Government budget allocations to supporting biomedical research, in millions of dollars (1985 prices).
Source: Data from [47].

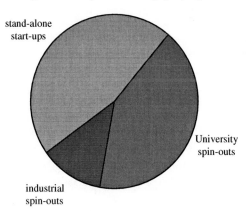

Origins of start-ups in Cambridge (UK) region

stand-alone start-ups

University spin-outs

industrial spin-outs

Figure 4.7   Origins of start-up companies
Note: Origins of start-up companies in the Cambridge area.
Source: Company list provided by Cambridge Enterprise, analysis by the author [69]. Figure copyright LGC Ltd 2008/reproduced by permission of LGC [69].

Nearly half the start-ups in this 'cluster' of companies around one of the world's leading research centres for biomedical sciences are not academic spin-outs at all – they were created *de novo* by entrepreneurs bringing intellectual property into their company from several sources.[4] US versus European state funding for biomedical R&D is irrelevant to this, as indeed is the number of papers. Genentech was founded from science spun out of one laboratory at the University of California, but soon outgrew this. Amgen was an aggregate of several sources of IP even at the start. Virtually all biotechnology companies follow this model – their intellectual property is drawn from the global IP pool, not from the university that happens to be down the road.

So the US does have the edge over Europe in terms of investment in biomedical R&D, and to the extent that it is easier to find and acquire IP within a continent as opposed to across the Atlantic this will benefit US companies. There are also better mechanisms in the US for providing government support to companies developing important new technology, and this has undoubtedly made a difference. The SBIR programme has provided substantial funds to companies developing some strategic new technologies, such as Affymetrix (now the world's leading gene chip company). By contrast European support has tended to focus on the creation of *companies* rather than the creation of *businesses*, with the concomitant result, noted above, that you get companies that are not sustainable businesses (I will discuss this further in Chapter 12.).

This, however, is not primarily about the quality of science but about investment in it, which brings us to the principle reason for the failure of European biotechnology compared to the US: money.

Having described why I think that the scientists, entrepreneurs, managers and markets are minor contributors to the equation, we are left with the investors, and in the context of early stage biotechnology companies primarily the VC investors. Even if all the previously mentioned causes did contribute to the underperformance of the European industry, it seems inevitable that investors must take some of the blame. As I discussed at the start, investors are intimately, some might say symbiotically, linked with the industry. The following six chapters will examine their contribution to the problems of the industry in Europe.

5
Underinvestment in European Biotechnology

Why do new companies fail? We have looked briefly at aspects of management and entrepreneurship, the usual scapegoats for company failure. However, the objective evidence is that overwhelmingly companies fail through undercapitalisation.

The crippling effect of the lack of capital on companies in general is well known. Riquelmo [89] reviewed ten studies of the causes of failure of small companies, and found that lack of initial capital was 1.5 times as likely to be cited as the cause of failure than the next most common cause, and that financial reasons were more than three times as common a cause of failure than lack of management experience. Compare Riquelmo et al's conclusions (summarised in Table 5.1) to the VC comments in management in Chapter 7.

Dimov and Shepherd [90] also found that by far the best predictor of success in a small start-up was the amount of money invested: this was far better than measures of 'human capital' (i.e. the management team strengths). The number of University spin-outs is also correlated with the availability of money [91], not with the size of the University or other factors discussed above.

Investor-backed biotech companies spend money in life science R&D to develop new products. This costs money. Without enough money, you only discover half a product, you only half-develop your drug and that is no use to anyone. Lack of money would be particularly critical to a biotechnology company. However, saying that a company died from lack of money is like saying a car-crash victim died from lack of blood. Giving drivers EPO[1] will not reduce road fatalities – the lack of blood is the end of a causal chain that started with driving too fast without a seat belt. Why do biotechnology companies lack money?

Table 5.1 Factors leading to company failure

Lack of initial capital	26
Inadequate or no accounting records	18
Poor credit management	14
Lack of management experience	17
Personal/personnel problems	16
Lack of planning	15
Lack of management and financial advice	14
Marketing deficiencies	13
Product deficiencies	13
Location	11
Economic conditions	15
Fraud	10

Source: Data from [89].

There could be two reasons: they started out with too little money or once started they generate too little money. This chapter will examine whether European companies that receive any investment receive too little. Note at the start that I am not arguing about whether there is too little investment in biotechnology in Europe. If there was 'only' £100 million invested in biotechnology companies in Europe but it was all invested in one company, then that company (with good management, good technology and without the problems I describe in Chapters 7, 8, 9 and 10) would be a world-beater. The issue is the amount invested per company, not a macroeconomic aggregate of activity across a country, and hence is also not about the *number* of VC houses in Europe but about their *behaviour*.

There is a persistent complaint from the UK biotech that investors do not invest enough to see an investee company achieve its goals. Investors deny this: they claim to give companies as much money as they need and deserve. It would seem foolish for an investor to provide less than the capital needed to add value to their investment. However, this is what seems to happen. We will ask why (and as before arguments that are *not* reasons why) soon, but because of the improbable nature of the assertion it is worth reviewing evidence to support it.

Several studies have suggested that European companies receive smaller amounts of investment at the same stage of their development as US ones: for example, the Bioindustry Association estimated that a 6 to 10-year-old company in US can raise 16 times more money than a comparable company in Europe [92]. The Critical I project [93–4] is one of the most comprehensive and recent such review. It estimates that US biotech companies receive 4.6 times as much private investment as European ones, and in the critical early years the difference is nearly eightfold (Figure 5.1).

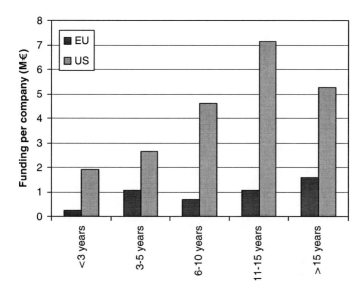

Figure 5.1 Amount invested in biotech by company age
Total amount invested in biotech companies of different ages.
Source: Figure compiled with permission from data kindly provided by Critical I, from [94].

Such studies are potentially flawed by sampling bias: based in the UK, they will 'see' a larger proportion of small, undercapitalised, low-profile UK companies than they will detect of US counterparts. In addition, Critical I was working at the depth of the 2002–3 slump in biotech investment, which could be considered unrepresentative. To avoid this problem, I surveyed all the companies that were funded by investors that regularly invest in UK biotechnology [95]. This included their US investments, and so should represent what these investors would put into any company that met their investment criteria, which will specify company size, 'stage of development' and needs for cash. The principal variable is therefore geography.

The results are shown in Figure 5.2. I have also characterised the companies by investment round rather than age. Conventionally investment in any company is done in stages [12], called the 'A', 'B', 'C', etc. 'rounds' of investment after the names of the share classes created at each round. The first major VC investment is the 'A' round. Figure 5.2 shows the breakdown of investment over the whole period 2002–5 by investment stage, and shows that for A and B rounds European, and particularly British, companies receive substantially lower amounts than North American ones. The amount appears more equal for D and

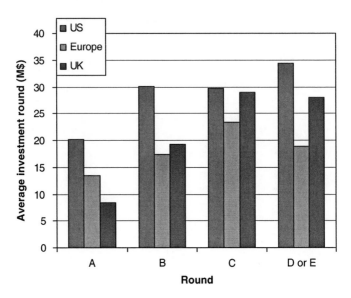

Figure 5.2   Investment in UK and US biotech companies
Average amount invested per round per company in US, UK, and rest of Europe, for 2002–5.
Note: Companies are those invested in by VC groups that have invested in >5 UK biotech-
nology companies in this time period.
Source: Reproduced from [95] with kind permission of Palgrave Macmillan.

subsequent rounds: this is, however, in part an artefact of the way companies are financed in the two regions, as investors rarely invest in D or
later rounds in European companies at all. A better reflection of the
amount invested in D and later stage rounds in Europe is that the
majority of companies have rounds of $0.

Over this period the number of investments also fell in Europe, but
kept constant (or fell far less) in the US. For example, Germany had three
VC investments in biotechnology companies in the first half of 2002,
compared to 70 in US companies: many of these US investments were by
syndicates that included European investors, showing that the failure to
invest in Europe was not because European VCs lacked cash [96].

Is this because European companies are inherently smaller than US
ones? This is unlikely, for two reasons. Firstly, as noted, these investors
have specific investment criteria: they will not invest in too-small
European companies (an issue that leads to the so-called *equity gap*, the
reality of which I will debate below). Ernst and Young have surveyed the
'survival time' of European companies, that is, the amount of their last
investment that they have left compared to the rate at which they are

spending it. This varies with age of the company (older, more secure companies have a longer runway) and date (in bull markets, companies have longer runways as they can raise more cash). US biotechnology companies have always had a longer 'runway' ahead of them than European ones of the same stage and date [27, 28, 93].

Secondly, it is also clear that the amount of investment per company has fallen since the start of the 'biotech era'. There is little systematic data on investment prior to 1998 (when the ready availability of internet search engines made compiling such data relatively easy), but an exhaustive survey of the Cambridge (UK) region, tracking companies back to 1981, has been conducted, and shows that the amount invested has fallen as the cluster of companies 'matured', and is now below the national average for Series A and B rounds (the East of England biotechnology industry forum data). In line with conclusions from Figure 6.2, amounts raised at IPO have fallen as well.

Thus, several different approaches to data collection confirm that European companies receive less funding than US ones that are at the same stage of growth or development, especially at the crucial early stages of the company's development. Individual comparisons confirm this, and also confirm that this underinvestment is endemic from company foundation, and hence are a cause, not an effect, of small company size and company underachievement in Europe. For example, Nereus Pharmaceuticals was a spin-out of the Scrips Institute (US), founded by scientists in 1999 to discover drugs from marine organisms. They received $8.6 million first round investment a year after foundation, which allowed them to recruit an experienced CEO, raise further finance to a total of $72 million, and drive two drug candidates through to clinical trials. Aquapharm Bio-Discovery was a spin-out of Heriot Watt University (Scotland), founded by scientists in 2000 to discover drugs from marine organisms. They received £50,000 funding in 2003, which meant that the founding scientists had to run the company and could take their lead compound no further than basic laboratory tests. They raised a total of £1.35 million (~$2 million at then prevailing exchange rates). As a direct consequence, their business model is to use their technical understanding to generate revenue and gain grants rather than focus on growth of larger commercial opportunities.

Other measures of how fast a company grows confirms that European companies are cash-starved compared to US ones. Two useful ones are patent filings and employment costs. To drive a biotech business forward, science must be done, and we would expect it to be the best science to compete on the world stage. This means that patents must be filed on key aspects of the products or technologies being developed, because no matter how excellent the science if it is not protected against copying then the commercial value in it will go to those with market muscle rather than to its inventors. Truly brilliant, innovative products require brilliant science and lots of hard work to realise, but the only barrier to pursuing a large portfolio of patents is the cost of drafting, filing and maintaining them. Pharmaceutical companies file hundreds of patents a year developing their products (Figure 5.3), and a biotech company is expected to follow them, albeit not at the same scale. Figure 5.3 illustrates that the major biotech companies do (the largest patent filer – Pfizer – is put on the graph for comparison – note that this is a logarithmic scale), and also illustrates that all the major European biotechnology companies together file rather less than each of the major US ones, in line with their size as reflected by market capitalisation (Table 2.1).

Lawrence [97] analysed patent filings in Europe and US in 2003, and while the number of companies was similar the number of patents filed

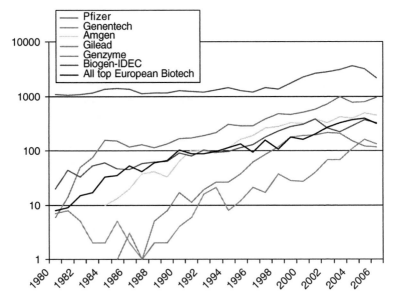

Figure 5.3   Pharmaceutical company patent filings
Note: Note that these are *published* patents – patents are published 1–2 years after filing, hence the apparent dip in patenting rates in 2005 and 2006. 'Top European companies' included are: Serono, Celltech, British Biotechnology/Vernalis, Basilea, Actelion, Peptide Therapeutics/Acambis, Innogenetics, Cambridge Antibody Technologies, Morphosys.
Source: Data from EU PTO, searched through Espacenet from fifth to seventh March 2007.

was 5–6 fold lower in Europe. (This is based on published patents – the patent process in Europe allows a patent to be *filed* very cheaply, but it is not published for 18 months, after which an average of £4k–£8k will have been spent on it, with substantially more to come. So these are the number of patents that someone has committed significant resource to.) Essentially the same figures are published in Capart [98]. This is exactly in line with the lower funding levels summarised above.

This is not as straightforward to interpret as it might appear. The US patent system is more friendly to early stage inventors, allowing a post-publication grace period to see whether the invention is any good, the Continuation in Part concept which means your prior patents do not act as prior art on your own improvements on your idea, and a common patent system (and single patent fee) for the whole US market [98]. However, as major market for many biotech products is US, this is not a serious issue, as any biotech companies with pretensions to being a global business would file in the US, and many chose to file *only* in the US, rather than in the UK. Capart [98] looking at IP differences between Europe and US explicitly finds that the principal differences between UK and US filing patterns are driven by good availability of investment in US.

(This leads to a bizarre conundrum. Investors demand patents, because they are the mark of a successful science-based company, but do not provide the investment to generate them.)

Another measure is the amount the companies can spend to recruit scientists. Although the biotech industry is at heart an investment industry (or more accurately, an industry generating raw materials for the investment industry), its business model, as I have described above, relies on scientists and technologists to invent new products to wow future investors. So science is central, and not just the 'routine' science that major pharmaceutical companies might be thought to pursue, but groundbreaking, paradigm-shattering science. Indeed, the best business model for biotechnology companies is either to develop a sales-based business based on what is known today or to pursue wild, 'off the wall' opportunities rather than to pursue the relatively well trodden paths that the mainstream pharmaceutical industry walks [10]. To do this, only the most creative, knowledgeable and hard-working scientists will do, and these people are in short supply. A consequence of market economics is therefore that good scientists cost more than less good ones, and you get the quality you pay for.

Despite one VC's claim that scientists in the UK are overpaid, 70% of R&D personnel in UK start-ups had salaries below £15k [99]. Only in sales and marketing did more than 25% of people have more than £25k salary. This is probably a fair reflection of their relative value to the business, but the absolute scale was a serious problem. Average salaries in US biotechnology companies at the same time was £33k–£39k, with scientists bringing in median range £26k–£31k, and senior scientists on a package of £33k–£38k [100] (dollar figures $50k–$60k, $40k–$48k, $51–$59k respectively). Scientists are a highly mobile population – working for two or three years as a 'postdoctoral' researcher in another country is a standard career move as we have noted from Figure 4.4 and Figure 4.5, (a 'postdoc' position is a fixed-term, salaried research position taken up after completing the PhD degree, and it usually lasts for two years). US labs were then and still are full of expatriate European junior scientists. If they are planning to move from these (usually academic) posts to a career in biotech, the attraction of at least double the salary will ensure that the best ones stay in the US. Regardless of this (there are attractions of living in the UK that might compensate for a much larger salary in a properly funded lab in the US), the salary gap illustrates an approach to funding to buy cheap with as little money as possible, rather than to buy the best.

Facilities set-up is also costly, and is underfunded in Europe. Kenny estimated in 1980 that it cost between $3.5 and $7 million to get a first-class molecular biology lab up and running for 2–3 years [18] pp 145–6. Setting up chemistry research facilities can be even more expensive. Adjusted for inflation to dollar values in 2000, this would be ~$6–$12 million. Anyone setting up a research group with less than this was therefore either setting up an incomplete facility or one that is second rate. The average amount invested in a Series A fund-raising in the UK recently has been around £2 million (Figure 5.2), which is well below this figure. In other words, the level of funding of UK biotech companies is setting them up to fail. The US figure is what is needed to set them up to succeed. 'It is catch 22 – the funders are reluctant to put money in because they have not seen great success in getting drugs through to Phase III, but unless the money gets put in you can't get there' (Glyn Edwards, CEO Antisoma, quoted in [101]).

And it is no point getting there second. The need for speed and the magnitude of first mover advantage is clear. To illustrate it, I will take two examples, one large, one small.

Mitchell [102] reviews the monoclonal antibodies (Mab) field. Mabs were one of the first types of biotech product, and remain one of the most successful. The first Mabs were fantastically profitable, the next wave less so mainly due to competition and the increasing difficulty of discovery. Once you have found the 'low hanging fruit', by definition what is left to find is harder. The market figures presented often did not reflect this, as many analysts continue to extrapolate the early success as if it can be replicated again and again. For example, Wiles predicted the Mab market to be $30 billion in 2006, but the top-selling six Mabs had a market share of $8.9 billion in 2005 (when none were off patent and all had growing markets), and with another six on market, eight in the process of being approved by the FDA and 150 in development the chances of another billion-dollar drug must be declining [103]. The race does not go to the cautious: it goes to the swift and the daring. In science, that takes money.

A smaller example is the ADME prediction field, which is not complicated by patent and legal costs, and as this example is not all that well known I will expand on it here. In mid 1990s ADME properties was the hot issue on which many drugs were failing [104]. (ADME stands for Absorption, Distribution, Metabolism and Excretion, and can be summarised as everything that happens to a drug molecule as it enters, passes through and leaves the body.) Getting the ADME right is critical – if a drug disappears from the body too fast, or never gets to the organ it is meant to treat or is metabolised to a toxic chemical, then it is useless as the basis of a medicine. Predicting ADME was therefore a major target, and a small number of specialist companies were created to provide such services to the pharmaceutical industry. Companies founded then (Camitro, Entelos) and well funded to grow fast and gain a market position were successful. Ones founded in the late 1990s which had to undergo two or three European-duration and size financing rounds before they had gained enough technological strength to be able to address the market's needs found that the market need had been met: by 2002 the leading problems leading to drug failure were not ADME but toxicology, and commercial issues around re-imbursement [104]. Established technology had found a place in the pharmaceutical industry and *solved the problem*. So the strategy (if it can be called that) of starting a company in 1998 but allowing it to grow so slowly that it could not be successful until 2002 inevitably resulted in failure.

The nature of company failure also points to underinvestment in Europe. 'Failure' in terms of company bankruptcies are rare in Europe, but failure in terms of major setbacks, radical restructuring and 'orderly' wind down of operations are common, as we noted in Figure 4.2. Failure could happen soon after the start (implying poor science), as the company moves from research to development (implying science that was not developed properly) or once the product moves to commercialisation (implying lack of business skills). A study on failure supports the second

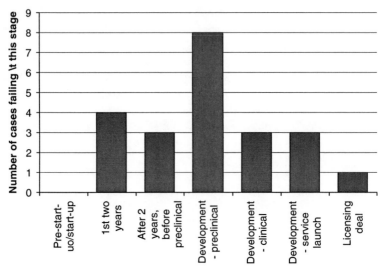

Figure 5.4   Failure in UK biotechnology start-ups
Note: Stage of company development at which substantial setbacks or outright failure happened, for UK early stage biotech companies.
Source: Figure copyright LGC Ltd 2008 reproduced by permission of LGC [69].

of these three alternatives. It showed that nearly half of funded UK biotech companies have substantial technical problems when they make the transition from research to development [69] (Figure 5.4). Research *can* be done on a shoestring, but the result is usually unreliable and cannot be taken into the more demanding and rigorous world of product development and industrialisation. Some of the companies concerned never recover, the rest waste up to 20% of their scarce cash trying to dig themselves out of the resulting hole.

The study emphasises that these problems could almost entirely be avoided by doing the science well in the first place – this is an entirely avoidable cause of failure, and due primarily to underinvestment [69].

In a highly competitive, fast-moving field, there are other reasons than scientific or technological ones to equip companies effectively. Having an effective IP strategy means defending your patent portfolio as well as drafting and filing it. Genentech had nine lawsuits running in 1988 [105], and had 12 lawyers in the legal department, six focusing exclusively on patents. This is driven by concerns to defend their market, not 'patents for patents sake': they were not fighting an alleged infringement

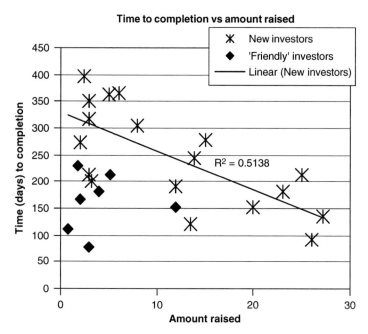

Figure 5.5 Time taken to invest in UK biotech companies

Note: Y axis – time (days). X axis – amount raised. Investments are separated into 'new investor' investment, where at least one investor was new to the company, and 'friendly investor' investments where all investors are familiar with the company. Solid line – least squares best fit linear line to 'new investors' category, with R^2 value.

Source: Data collected by the author.

on their patent on gamma interferon (on which they claimed priority) because they did not consider market large enough [106]. By the mid 1990s Affymetrix was fighting five lawsuits with three companies on who owned the rights to basic aspects of their technology [107]. (These were all settled in the end, but one informed estimate put the aggregate legal costs of all the parties at $40 million.) If you do not have the financial muscle behind you to look credible in IP defence at this level, there is little point in filing patents in the first place.

As well as the failure to provide companies with enough funds to be a success, Europeans also take an inordinate amount of time to provide them with anything at all. This stretches company timelines, wastes executive time and enervates the entire entrepreneurial team. I surveyed nine entrepreneurs and investors with personal knowledge of a total of 27 recent (post 2001) VC investments in UK biotechnology companies. The results are summarised in Figure 5.5.

The time taken to invest was very varied, as you would expect, depending on the fund involved, the VC management group, the company management team etc. However, 40% of the variance can be explained simply by separating 'new' investors from 'friendly' investors, and assuming the time taken to invest for 'new' investors was inversely related to the amount invested. In this context, 'new' investors are ones which are not already intimately familiar with the company, that is, ones who have to gain familiarity with the management. 'Friendly' investment rounds are ones where *all* the investors are already familiar with the company, either because they worked for it or because they have invested in it. The second of these is the most common – the 'Friendly' rounds therefore primarily represent closed rounds, usually top-up rounds or minor scale refinancings. For the 'non-friendly' investment rounds (i.e. ones where at least one, but not necessarily more than one, of the investors was new to the company), the average time taken to raise £10 million or less was around ten months from the time the first business plans were sent out to closing the deal. This data set represents over ten different VC groups investing at the peak and the trough of the economic cycle, in many different businesses with different sizes, expected times to exit and different products. But the message is remarkably consistent.

While this might be understandable in terms of the VC processes, due diligence requirements, deal flow handling and many other procedural issues, it is a crippling burden on companies. Typical early stage VC funding only provides a company with 1–1.5 years' of operating costs [108]. This is not a new phenomenon – Gompers comments that in the 1990s VCs usually gave companies about one years' money, [108] ($1 million–$2.5 million as a first round, based on data from 794 companies 1961–92). In general the time bought by an investment actually declines as the company develops (1.6 years' funds at seed stage to 0.86 at third development stage), although the absolute amount goes up. This is particularly notable for their investment in high-technology companies (in part because low tech companies accumulate physical assets which the VC can recover if the business fails, while high-tech ones accumulate IP which is almost valueless outside the context of the company and programmes that created it). Both BIA and Critical I [93] commented that around 30% of UK biotech companies were so poorly capitalised that they had less than a year's funding. If it takes nearly a year to raise further funding, it is clear that the company effectively does nothing but fund-raising.

I mentioned above that the length of time taken to invest might be explicable in terms of the investment process, and particularly of the 'due diligence' evaluation that investors undertake of potential investee companies. In major public financing 'due diligence' means a thorough and systematic verification of the legal and financial status of the company, including stock and sales records and of the status and standing of all the parties involved with it. In the case of a private biotech company, and especially an early stage company, much of this is irrelevant, and 'due diligence' blurs into the 'opportunity evaluation'. The 'diligence' aspects come from verification that the science and technology is as the company portrays it, that their market figures are not provably ridiculous, that the founders are not going to bring the company into discredit and so on. The actual rigour of this process has generally not been examined, because it is not in investors interests to do so, but one study of the due diligence process [69] found that the level of 'diligence' used was in nearly all cases lamentably low (Figure 5.6).

Seeing these data, most VCs have commented that it is only to be expected that they would not involve themselves, even through intermediaries in the specifics of the science or technology, in the overview.

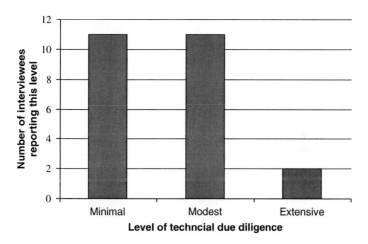

Figure 5.6 Technical due diligence levels in start-up biotech fundings
Note: The level of due diligence performed on the scientific or technical base of start-up and early stage (post start-up) investments in UK biotechnology companies, 1998–2005. Due diligence is 'Minimal' – due diligence is confined to taking references on the scientists involved, reading their papers and a summary report from the company. 'Modest' – in addition to the 'Minimal' level, external experts are brought in to receive summary presentations on aspects of the science and to interview key scientists or 'Extensive' – in addition to 'Modest' level, investors or how their agents view methods, Standard Operating Procedures (SOPs), laboratory processes and primary data reports.
Source: Figure copyright LGC Ltd 2008 reproduced by permission of LGC [69].

Such comments only serve to reinforce Figure 5.6 and illustrate that, whatever takes 320 days in the investment process, is not receiving a 90-minute PowerPoint presentation and a bunch of CVs from the company's founding scientists.

VCs also typically 'stage' investments. A commitment to invest £2 million might be split into two stages, so the terms of investment are agreed for both stages at the start but the second 'tranche' of cash is only given after the company achieves specific milestones. This staging mechanism is a control mechanism for the VC – companies are constrained to follow the path the VC has laid out for them, or they do not get the rest of the cash. Negotiating with the VC about whether those milestones have been passed, especially if the business has changed and they are no longer relevant to business success, itself takes time. Sweeting found that two out of four funds interviewed refuse to enter deals with staged payments against milestones because the 'renegotiation' that goes on when the stage is due is too time-consuming and often not resolved to anyone's satisfaction [109].

So the combination of drawn-out fund-raising processes and small amounts of investment at the end mean that nearly all of the top management's time can be expected to be taken up by fund-raising. The company will then get enough money to take it forward another year, all of which will again be taken up by fund-raising. Thus, the length of time taken to distribute the funds renders them even less valuable than their monetary value would indicate. In the extreme case of a new start-up with a management team of only a couple of people, typical UK funding levels are sufficient solely to allow the company to seek more funding, never to actually achieve anything.

Biotech research, and drug discovery in particular, is a race. Car races are not won by low-power, cheap cars: they are won by the best cars and the best teams to drive them. A Citroen 2CV might be fine for shopping, but it will not win the Grand Prix. Fast commitment and substantial funding are essential. Slow commitment of small amounts of funds will produce the 2CV of biotech companies.

There is an explanation for this apparently catastrophic underfunding of early stage European companies, which can be summarised by the phrase 'equity gap'. Different investors fund companies at different stages. 'Business Angels' typically invest at the earliest stages of a company's history to get the company off the ground. Early stage VCs then invest in the established company, VCs specialising in development capital invest at later stage, investment banks provide mezzanine capital prior to flotation and then the public markets finance the company into profitability. If at any stage there is a range of funding which is not covered by these funding mechanisms, then there is an 'equity gap', and companies will be stranded on the low-value, low-investment side of this gap, unable to grow further to meet the requirements of larger, later-stage investors.

Much has been made of the 'equity gap' in Europe, and particularly the idea that private investors – business angels – cannot provide enough funds to allow companies to develop to sufficient maturity to be suitable for VC investment. In Europe and the US business angels support far more companies than VCs. Mason estimates that 85% of companies are angel, not VC funded in the UK [9]. Reichhardt suggests a similar figure for the US, and estimates that they invested a total of $22 billion in 2004, 10% of it in biotech [110]. However, in Europe (the argument goes) angels do not support companies to a sufficient level of maturity to allow VCs to take them on. This is why VCs do not invest. So developing companies are forced to return to their existing investor base, which is equipped to invest only small amounts, for further funding, rather than stepping up to the more developed and much larger VCs for substantial funds.

It is certainly true that it is relatively easy to get 'seed' funding from a variety of government-associated sources in Europe if you apply at the right place and time (as I discuss below in Chapter 12). Understandably, VCs do not want to invest in many of these companies, which are in reality the hobby of the founding scientist. But beyond this the 'equity gap' argument does not hold up. VCs in Europe invest £500k and up, depending on the fund, and have done since 1995. Angels invest (as consortia) up to £1.2 million in start-up biotechnology companies, and have done so since 1995. So apart from the lack of funds willing to prop up uncommercial, state-funded 'spin-outs', where is the equity gap?

Trying to tie down where this gap is from empirical surveys of companies' fund-raising problems suggests that it is not so much a gap as a canyon. Thus, in 2004–5, for example, different studies said that 'the equity gap' affected biotechnology companies in getting seed funding (i.e. the first funding of up to £500k) [111], companies that had raised £250k and were looking for £1–2 million [112], companies in product development pre-Phase II clinical trials (i.e. companies that had raised £2 million – £10 million and were looking for an additional £5 million) [113], companies that were 'late stage' and ready for IPO (i.e. had received at least £10 million in several funding rounds) [114]. In short, the 'equity gap' is the financing continuum between angel investment and public markets, or to put it differently, the equity gap is the VC investment market.

This has been a problem in Europe for at least a decade. Murray [115] describes a major 'equity gap' for start-up finance, and for funding between companies at this stage and 'development capital' organisations. Indeed, the Macmillian report in 1931 [80] suggested that firms would have difficulty raising amounts less than £20k, ~£4 million today, that is, the median of the investment amount that VC now claims to provide. In recent years companies have been so frustrated by the 'equity gap' that 'Technology Management Companies' [116] have arisen which fund biotech start-ups from Angel funding or their own resources and then float them as micro-cap companies on public markets, notably the London Alternative Investment Market (AIM). This is often stated to be 'filling the equity gap' but in fact does so by avoiding the VC funding route [117], rather than by filling in a gap between start-up funding and VC. Other studies have confirmed that the 'equity gap' is not a reality for companies seeking investment – mechanisms and sources exist for all stages of investment, and the primary problem is the unwillingness of VC to invest at any stage, amount or price [118].

A related concept is that of 'investor readiness'. Harding [119] used the Global Entrepreneurship Monitor (GEM) to analyse scale and nature of investment in UK. He found an oversupply of capital alongside inability of some companies to access capital, and concluded that they are not 'investor ready'. Companies could get investment if they were 'investor ready'. My attempts to tie this concept down with the many groups advertising 'investor readiness' services (for a fee, never on success payments) identify the core concept in 'investor readiness' in 'making the company seem more mature than it is'. Critical components usually include financial modelling for five -year accounts forecasts, making the business plan look 'professional' (usually by removing any useful description of the technology), refining the business model to suit what investors want to hear and so on. The pointlessness of this, especially five -year financial forecasts for businesses that are reliant on year-on-year investment from the very people those forecasts are sold to, should need no further exposition here.

In the US, early stage investors are quite explicit about *not* requiring such a 'gloss' to be put on companies before they invest. For example, Versant's presentation to Bio2E (http://www.bioe2e.org) in 2001 listed six of their investment success stories – CV Therapeutics, Aviron, Tularik, Onyx Pharmaceuticals, Symyx and Athena Neuroscience – that had exited through IPO and had a market capitalisation (at that time – mid 2001) of $5 billion. None of them had approached Versant with a 'business plan'. The reality of early stage businesses is that a 'business plan' is of little use. Less than half, growth-orientated owner-managers use a business plan [120]. And business plans for growth are usually almost completely uncorrelated with what happens. Having a sophisticated, spreadsheet-rich business plan is not a requirement for running a business or for success or for gaining investment from experienced VCs. It is a reflection that in order to gain investment of any sort in Europe you must appear as a developed, low risk opportunity. This is clearly not appropriate for a biotech start-up, and not what most people understand that 'venture' capital invests in.

So 'the equity gap' is in fact an illusion. Raising cash is always hard – the demand for money will always outstrip its supply. But the evidence of empirical surveys shows that there is no single 'gap'. Companies anywhere along the development path, looking for any level of funding, will find that the 'gap' has moved under their feet if they are looking for VC funds. 'The equity gap' is just another way of describing VC underfunding of European biotechnology companies.

Before I launch into a harangue on how none of the reasons commonly given for this crippling underinvestment are defensible, I must comment that entrepreneurs are to blame, often as much as the investors. For reasons we will discuss (and false arguments that we will dismiss), VCs want to keep investee companies on 'short commons'. But entrepreneurs also want to take as little as possible. As Gorman and Sahlman concisely comment [121],

> [e]ntrepreneurs, because they are motivated to retain for themselves as much as they can of their business' value, are loathe to incur any up-front expenditure that might conceivably be avoided. Selling too much of the company at the earliest stages amounts to an expensive mortgage on the future wealth against which the entrepreneur has wagered his or her career, Thus both venture capitalists and entrepreneurs willingly conspire to impose stringent limits on the resiliency of their enterprise.

Why do entrepreneurs not realise that accepting more investment could build a bigger, more robust, much more valuable company? Many do, but find that they cannot obtain the funds anyway. At an investors' conference in 2007, I watched a promising UK start-up ask the investor audience for £450,000 when their plan clearly needed £5 million and the opportunity justified it. Why not ask for more? The CEO shrugged – there was no point. No one would give them £5 million, so you ask for what is attainable. However, not all entrepreneurs are so unambitious, and there are many who would ask for what they need to grow their business fast and effectively. The next two chapters will examine some of the reasons given for underinvestment, and why they are not valid excuses for VC actions.

6
Public Markets and Underinvestment

VC exists because it promises to invest in bright new companies and give them the cash to grow, thus generating share price rise and the chance of substantial profit on their investment. The sections above show (in rather tedious detail) that it does not do so, and that the 'gap' in finance in Europe is in fact a reflection of this. I have argued that this is not because companies are not there to invest in, nor that the science, management, entrepreneurial drive or other features are not ready to feed the VC machine with raw stock. I have also touched on how under-investment increases biotech companies' chances of failure. This is not because European VCs have less funds than US ones: fund sizes across all stages of the fund life cycle are similar [17]. So why should VCs pursue this policy?

By far the most common opinion I have heard expressed, once one has dealt with the fallacies of the defects of management, science and so on, is that this reflects the poor public capital markets in Europe.

To recap on some basics, when an investor buys shares in a company, the critical question he or she asks is what the price of the share, and the implied valuation of the whole company, is. This is just as true for public fund-raisings as for private ones, including the 'Initial Public Offering' (IPO) when the company first offers its shares on a public stock market. For public markets, however, there is an additional complication. People with shares in a private company are usually constrained by the rules of the company from selling those shares – this is understood by the investors. People with shares in a public company expect to be able to sell them when they chose. However, this means that someone else has to buy them – this is share trading, not the company issuing new shares (except in the specific cases when a company does issue new shares to raise more capital). If a pool of buyers and sellers reflects the stocks' 'liquidity' (i.e. how easily the shares flow around the market) – if it is illiquid, then no matter what the theoretical value of your shares is you cannot sell them, and they are of no practical value to you. Both valuations and liquidity are a major issue for biotech companies.

There are a number of other issues around being a company whose shares are publicly traded (a 'public company' for short) that arouse the ire of biotech investors. Complaints of poor IPO valuations, low liquidity, risk-aversion and legal constraints such as over-regulation, lack of a NASDAQ equivalent, pre-emption rights, inability to raise further funds in follow ons, the ignorance of the analysts working on public market biotech stocks and others have all been cited. Remember that the biotechnology industry is *not* a product or science industry – it is an industry generating raw materials for an investment industry. If it is not possible to sell the shares that VC has acquired in a company for a good return on investment, then of course the VC would be foolish to invest in that company. More quantitatively, the VC should only invest that amount that, when the company is floated or sold, will yield a good return. If European IPOs never go for a pre-money valuation of greater than £10 million, then no VC should invest more than £3 million in any European biotechnology company: if they do so, there is no hope even in principle of getting the 30%–50% annual Return on Investment (ROI) that they require to make their business work.

In fact, VCs do not do this calculation, for reasons I discuss in Chapter 11 below, but here I wish to argue against this thesis as if VCs really did do this calculation. More detail on this has been published in [95]: this section summarises that data, and extends it with some other studies.

The 'good exit' issue is central. It is stated that European stock markets will not support high value IPOs, so there is no point investing substantial sums in companies before IPO. This is fallacious for two reasons.

Firstly, it is quite clear that it is possible to get very high value biotechnology company IPOs in Europe. European markets, and particularly the London Stock Exchange, can accommodate floatations valuing European biotechnology or emerging pharmaceutical companies at over $1,000 million. Markets will accept high-tech story for a good company. However, it is true that the average IPO pre-money valuation is less in Europe than the US: in US biotechnology companies have twice the pre-money IPO value and raise twice as much money (Figure 6.1).

But the money was there: in 2005, despite there being few companies of any size, European biotech companies raised more in IPOs than US ones [122], although the average cash raised per company was lower. The argument that 'European markets do not have the appetite or

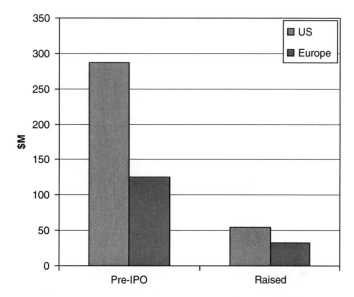

Figure 6.1 Average pre-IPO valuation and average cash raised, US vs EU
Note: Average pre-money IPO valuations of biotech companies founded between 1995–2005, and amount of money raised at IPO.
Source: Reproduced from [95] with kind permission of Palgrave Macmillan.

capacity for substantial biotech investments' is untrue as a generality, although at specific times it can be cripplingly true (in Spring 2008, for example, biotech stocks tumbled because there was absolutely no appetite for them at all). This is illustrated by the Technology Management Company (also sometimes called the Technology Exploitation or Technology Commercialisation Company) [116, 123] such as IP Group and BioFusion, which raised funds to invest in very early stage concepts, often concepts that have not been captured into a company at all. These have been a signal success on the UK AIM stock exchange, a success that shows that there is a greater appetite for high technology, early stage companies on the public markets in the UK than there is among VCs in the UK [123].

It is also untrue that US IPOs all value companies highly. Genentech's IPO was famously frantic, the stock price doubling in the first hours of trading and valuing the company at $400 million. But subsequent IPOs in the 1980s did not follow this trend: they raised an average of around $30 million and valued companies at an average of $100 million, around £60 million–£70 million at the then prevailing exchange rates, which is in the mid-range for European companies [124]. Only during the genomics stock 'bubble' of 1999–2000 were huge US valuations seen routinely [124–5]. Comparing a routine European IPO to these exceptionally successful examples is not comparing like-for-like.

High value trade sales are also possible. Medisense sold for $876 million, and in recent 2005–6 KuDoS was acquired for £120 million, Arakis for £106 million, Domantis for £230 million and Solexa reversed into Lynx valuing the company at £180 million. None of these are small figures, and are higher than average for US companies: US IPOs in 2005 had a median pre-money valuation of $165 million (~£90 million), which fell to $110 million (~£58 million) in 2006 [126]. So the European markets do value some European companies highly: there is *not* a structural 'glass ceiling' in European public market valuations.

There could be two reasons for European IPOs being lower in value than US ones. The first is that equally good companies could come to the markets, and the European markets could value them lower than US markets. This is the implication of the argument that 'you cannot get good IPOs in Europe'. Alternatively, European private investors could be bringing poorer companies to the market, and both European and US markets value them objectively according to a global standard.

The data suggests that the latter is the case. Figure 6.2 plots the amount invested in biotech companies *before* they IPOed against the value of the company at IPO. There is a clear ($R^2 = 0.48$) correlation between the amount of money invested in a company in its pre-IPO history and its IPO value. The relationship for US and European companies is essentially the same, and European companies that receive substantial investment can achieve high IPO valuations well within the US range. Investing more, in line with the arguments in Chapter 5 above, builds better companies. Similar comparisons done by a European VC house have come to the same conclusion[1].

Gompers [108] finds a similar correlation in a US context – more money in means a more valuable company. McGully also has noted a similar effect for US companies, finding that the 'technology value' of a company at IPO is on average equal to invested cash [127], i.e. that investing a lot creates a lot of value, investing little creates little value. This is in line with the careful and realistic evaluation of new IPOs by public analysts in the market window of 2005–6 [128]. As noted in Figure 5.2, European companies usually receive less money than US ones. Lower IPO prices are inevitable. In essence, if you put a lot in, you get a lot out.

I noted above that the amount invested historically in early stage Cambridge (UK) companies has fallen, and the amount raised at IPO

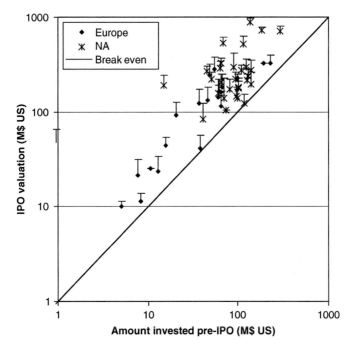

Figure 6.2 US and European biotech IPOs 1995–2002
Note: Shown are the amounts invested by private investors (X axis), versus pre-money valuation at IPO (Y axis). Y axis bars are the amount raised at IPO.
Source: Reproduced from [95] with kind permission of Palgrave Macmillan.

has also fallen. Figure 6.2 suggests that this is a causal effect – building less valuable companies through lower investment has resulted in less valuable IPOs. The amount invested at later stages has risen over this time period – however, this has not resulted in higher value IPOs. Propping up companies with continued investment does not compensate for the failure to capitalise them properly from the start.

The same case can be made from the perspective of companies as vehicles for business generation. For a class of biotech companies now out of fashion, called 'technology platform companies', their product was not meant to be a drug but a technology that could be sold to other companies. The overwhelmingly most successful of these companies in financial terms in the decade 1990–2000 were the genomics companies. The measure of success of such companies was their stock price, but also the number of major collaborative contracts ('deals') they did with pharmaceutical company partners to apply their technology. This is a

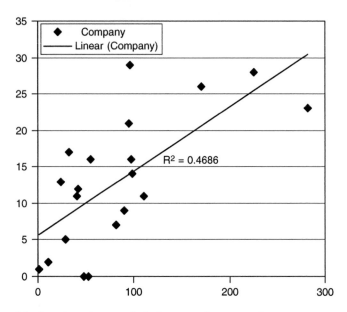

Figure 6.3   Investment versus deals for genomics companies

Note: Number of deals signed by selected genomics companies versus investment in those companies in the first six rounds of investment (from any source). Y axis: number of deals announced. X axis: investment ($ million). Solid line – least squares linear fit to data, with R^2 value. Data for: Abgenix, Cambridge Antibody Technology, Exelexis, Gene Logic, Genetics Institute, Genome Therapeutics, Genomyx, Genset, Human Genome Sciences, HySeq, IG Laboratories, Incyte, Integrated Genetics, Lexicon Genetics, Lynx Therapeutics, Mercator Genetics, Millennium, Myriad, Oxford Glycosciences, Paradigm Genetics, Sequana Therapeutics, Vysis.

Source: Copyright © (2000) Society of Chemical Industry. Reproduced with permission from reference 129. Permission is granted by John Wiley & Sons Ltd on behalf of the SCI.

different measure to IPO capitalisation, and one more attuned with the real business world of making money.

There has been much complaint that Europe did not produce a genomics company to rival Millennium or Incyte in the US either in deals or in market capitalisation. But the reason is clear. If the number of deals is plotted against company size, there is a strong positive relationship: having more to invest in technology means having more technology to sell, and hence having more deals to do. This has been done for data from the start of 2000 in Figure 6.3.

Investing in top science, top facilities and rapid growth delivers a rapidly growing, world-class company, which attracts investment from other companies and from investors.

Of course, I predicted at the time that the genomics company business model was unsustainable [129], and this has subsequently turned out to be the case. However, from a shareholder point of view this does not matter: the share price in these companies rose, remained high for some time, and so provided the opportunity for substantial returns on investment. And the businesses themselves generated revenues, paid salaries and discovered products. If the aim is to do any of these things, then again this illustrates that investing small amounts builds small companies and correspondingly generates small returns. European VC can only blame itself for not having a European Millennium or Incyte – the raw feedstock was there, but ended up supporting much smaller endeavours such as Oxagen, Hexagen or Gemini which, if they were lucky, were then bought out by their larger US rivals.

This does not mean that investing large amounts guarantees a good IPO, or good post-IPO performance of the company, as the recent history of several companies attests. Poor business ideas and poor management will kill a good company no matter how much money it receives. But it does show that, in biotech as elsewhere, inadequate capitalisation leads to small, inadequate companies.

If the returns on investment in European companies, either in terms of IPO valuations or in terms of major investment from pharmaceutical partner companies, are what you would expect from the amount invested in them, why do European investors think that their IPOs are poor?

It is possible that the same absolute return for European investors is not as valuable to them as for US investors, because Europeans take longer to invest than Americans, so the annualised returns are less even though the absolute return is equivalent. Figure 6.4 suggests this is not so: the time taken from initial investment to IPO is similar for US (average = 5 years) and European (average = 5.5 years) companies. (I have commented before [22] that the average time for *all* European biotech companies to IPO is 11 years: however, this includes a number of companies that are run for decades as family or self-funded concerns before going to public markets. We are concerned here with the history of companies that receive external investment with the goal of realising investors share value through IPO at the earliest opportunity.) Of course, more of the 5.5 years is spent just getting the money, so the funds are not used so effectively, but the time factor in the ROI calculation is not an issue.

We should note here that there is a cultural problem in Europe relating to timing. While the time taken to get a company from start-up to IPO is similar in US and Europe, the speed with which institutional investors capitalise on market sentiment to get a company to IPO is different. The IPO peaks of 2000–1 and 2004–5 started in the US, and were followed by IPO surges in London 6–9 months later [130]. This will affect the amount that companies floated by European investors on

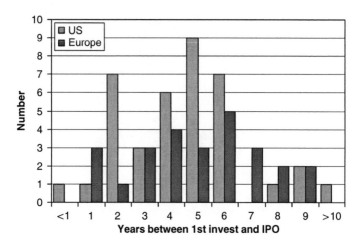

Figure 6.4 Time to IPO for European and US biotech companies
Note: Number of years between first VC investment and IPO for 64 US and European biotechnology companies.
Source: Reproduced from [129] with kind permission of Palgrave Macmillan.

European stock exchanges can raise. At the start of a stock 'boom', even a minor one, investors are interested in the field and the future – by definition, by the end of it they feel they have enough exposure to the future, and do not want to invest more. As stock markets are global, the market that gets its companies out first will therefore raise more money for them. This is not primarily a failing of the floated companies, however, as access to stock markets is primarily controlled by the financial advisors, backers and underwriters involved in the IPO. So if these institutions feel that they are missing out to the US, the solution is within their own control, and is one of speed and efficiency, not investment quantum or the nature of the business concerned.

It is also possible that VCs hope to get US-style IPO valuations for European-style investment. This would be very attractive financially if it could be achieved, but Figure 6.2 suggests that it rarely can, and for the very good reason that you cannot build a world-beating company on the cheap, as I have discussed. In any case, the 'US-style IPOs' are themselves very rare, even in the US: most IPOs in the US are not headline-busting, billion-dollar floats, but much smaller affairs as the discussion and data above attest. In reality, European investors are comparing today's European IPOs with the headline-grabbing billion dollar IPOs that a few US companies have achieved, not with the average.

This also illustrates why arguments that European governments must step in to help investors get good exits, made in Germany [131] and the UK [132] in 2003 are nonsense. In reality there were lots of exits in 2003, the period when companies 'could not get exits', as trade sales and as IPOs. Very few of them, however, were over-priced and under-funded VC-backed companies.

A related reason presented for poor IPO performance in Europe is that there is little appetite or capital for follow-on financings in Europe. In the US, the argument goes, an IPO is only a start. The company can come back to the market again and again, asking for huge amounts for quite general aims, allowing it to build vast war chests of cash to grow or acquire. This is not possible in Europe. So investors say (with some validity) that there is no point investing in an IPO stock if the company is doomed to fail for lack of follow-on finance.

Where did this idea come from? There are two sources, neither of which stands scrutiny.

Firstly, it is believed that there may be more mechanisms for invest-ment in US public companies as compared to UK ones, allowing contin-ued growth after IPO. As well as follow-on fundings, the US markets have been creative in developing biotech-specific vehicles for investment in public, mid-cap companies, from the Special Purpose Investment vehicles of the 1980s to the programme-specific investments made by Symphony Partners and others in the 2000s [133].

In particular, PIPEs (Private Investment in Public Equity) are an accepted mechanism of allowing private equity groups (i.e. VCs and similar invest-ment vehicles, as opposed to large, public investment groups such as mutual or pension funds) to invest in public companies in the US. It is often stated that PIPEs mechanisms have not been put in place in Europe: this is, however, incorrect for two reasons. Firstly, there is no unique 'mechanism' involved – no laws need to be changed or governments top-pled to make these things happen, and in fact in the late 1990s and early 2000s PIPEs were used as an investment mechanism in Europe, although not called by that acronym (see, for example, [94]). Secondly, even if rules or regulations did need to be changed, they can be changed: the barrier to doing so when the investment community wishes the rules changed is not high. When the investment community wished the rules to be changed in the UK so that loss-making biotech companies could be floated, the 'Chapter 20' rules allowing the flotation of 'science research-based com-panies' were introduced relatively quickly to allow this to happen. At most, the lack of PIPEs-like mechanisms in Europe is another reflection of investors' lack of interest in continuing to support such companies.

But secondly, it is believed that European markets will not allow companies to continue to raise substantial funds after an IPO, while US ones will. The Cooksey report is typical, commenting in 2003 that US companies raised $6.4 billion in 2002 in secondary public offerings, whereas UK ones raised £100 million, and concluding that there was something structurally different (and by implication, wrong) with UK public investors that caused this [134]. The contrast to the US is typical: the following is not an exact quote, but is an amalgamation of several people's comments to me, and is typical of comments that several funding professionals have made to me:

> In the US you can go back to the markets for substantial sums to build your company after IPO. In 2000, Celera raised nearly $1 billion in a secondary offering only 'for general business purposes' just 6 months after their IPO! That would never happen in Europe. Analysts do not understand the need to grow companies: they do not want to see you back to the market for more cash for at least 2 years after the IPO, and then will only allow you to raise small amounts for very defined purposes. They do not understand technology, they do not understand risk, and so you just cannot raise the cash to grow another Amgen in Europe. This drives European companies to small-scale programmes, small-scale ambitions, and usually to sell out before they can become product-generating, profitable companies.

There are three fallacies in this argument. Firstly, as we have seen, it is quite possible to raise large amounts on European markets *if* you have a convincing company in which the markets can invest. But floating a £10 million company and then expecting analysts to hand you half a billion pounds is clearly silly. Related to this is the belief that in the US, industry raises a majority of its finance from public markets: this is just not so, and biotech companies in the US are as dependent on VC as they are in Europe. Although it is true that US companies raise a slightly higher proportion of their financing on public markets than European ones (Figure 6.5), that is, probably because they were bigger and better established companies in the first place.

Secondly, it is assumed that the public markets are ignorant and conservative, and will not support a £10 million company at a valuation of £10 million, because of fear of innovation. Whether they are ignorant or not is hard to judge, but they are not inherently more conservative than VC – if anything, rather less so, as illustrated by the Technology Management Companies mentioned above [116, 123], a company created

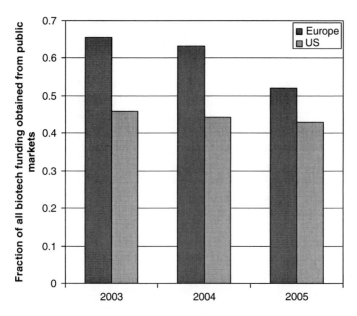

Figure 6.5   Ratio public to private sources of finance
Note: Ratio of amount raised by biotechnology companies on public markets versus the amount raised in private markets in Europe and the US.
Source: [135].

and floated on public markets with a specific remit to invest in the very earliest stages of high-tech company creation. IP Group (formerly IP2IPO) was the leader in this field, floating on the AIM market in 2003 and seeing its stock double in price over the following nine months (and not fall substantially after up to the end of 2007). IP Group invested solely in very early stage University spin-out companies, defining their market as those companies conventional VC would not back: despite this, their stock price was consistently bullish. Such groups show that there is an appetite for high risk, speculative companies in high-tech Europe. Figure 3.4 also illustrates this.

Thirdly, the comparator is flawed. The huge amounts that are cited as being raised by US companies were all raised during late 1999 or early 2000. This was effectively a stock bubble, which lasted for about six months and resulted in some enormous fund-raising by companies whose controlling shareholders were fast enough to exploit it, and whose companies were mature enough to look credible in the light of the market expectations of the time [124]. Examples include Celera (raising

$938 million in a secondary placing, which it stated that it would use for 'general corporate purposes, including possible acquisitions, alliances or collaborations' among other reasons), Maxygen (raising $146 million in a secondary in March 2000 after achieving an IPO in December 1999), Medarex ($359 million), GeneLogic ($270 million), Celgene ($251 million), Cell Genesys ($176 million) and Maxim Pharmaceuticals ($176 million) [125]. The biotech industry raised $12 billion in the first quarter of 2000, more than four times the comparable period the year before and far more than any quarter since [125], virtually all in the US.

How did this happen? All these examples are large, established companies (compared to Europe, anyway) with products in development, that is, they are companies that, with 'only' a few hundred million more investment, were considered likely to break into profit. As we have mentioned before, this expectation turned out to be unrealistic, but it was nevertheless there. Before the biotech stock bubble of 2000 their stock prices were already rising as investors realised that they were nearing product-driven business models based on unique, visionary science that would give them a competitive edge [136]. They are not early stage companies doing research that were pushed onto the public markets before their time. European companies could have dipped into the investor pool then as well if they had either been as substantial and established, or they had the vision and *cojones* or both. A very few did: the rest missed out. Since then US public markets have been nearly as hostile to small biotech companies trying to raise large amounts of cash as European ones.

In reality, European markets also respond well to ambitious, well funded, near-market opportunities. This observation – that in late 1999/ early 2000 US companies could raise lots of money while European ones could not and may never be able to – is not merely a result of underinvestment in the European companies in the first place. It is a strong confirmation that the public markets are *not* stupid, short-sighted or ill-informed in Europe, or at least no more so than in the US. They respond to a strong commercial story. If the earlier investors, and specifically the VC investors, cannot provide them with one, the public markets are not to blame. One analyst commented (privately) to me in early 2008 that it is not surprising that a company that had received £120 M from the public markets and is now worth £12 million cannot get investors interested in buying its stock. Investors are not stupid. They know a failure when they see one.

In fact, there are good reasons to believe that European public companies would fare worse in a US-style market than they do in European ones. In Europe fund managers who want to be 'in biotech' have limited options – they chose the companies that are floated on their native exchange or on one of the major European exchanges. US fund managers have the wealth of US companies to select from, which are better capitalised, more liquid and more successful. Translated into the US environment, even relative European success stories such as Nicox, GenMab, Biovitrum, Celltech (as it was until it was acquired) look unattractive to fund managers with billions to invest and limited attention spans for yet another $10 million struggling discovery-stage company when compared to US investment opportunities. Were the US and European markets to be merged into a 'global NASDAQ', observers believe Europe would probably suffer, not benefit [137], a direct result of their IPO market capitalisation which, as we have seen, is itself a result of financial starvation at inception.

In conclusion, there certainly are issues around capital market structures in Europe which could make it harder for European companies to raise finance. But many of these track back to the small size and low quality of those companies, not to inherent structural issues of the markets themselves.

7

VC Management of Companies

Not all businesses require VC investment – indeed, few do, which is a good thing as otherwise hardly any new businesses would be created. 'Normal' business is dependent not on start-up capital but on revenue, and specifically on cash reserves built up from profit to buffer revenue fluctuations and provide capital for growth.

Biotechnology needs much more start-up capital than most industries (possibly it is only surpassed by major aircraft manufacturers, nuclear power station builders and US presidential campaigns). But the same logic should apply to them as to a corner flower shop – keep money coming in and stop money going out. If a biotech runs out of money then surely the management has done something wrong. Even if they promised the VC a cure for blindness and the VC only gave them enough to invent a new spectacle frame, they should cut their budget to suit their cash, invent the new spectacle frame and go out and sell it. No one in business has an infinite budget, everyone has to sell. Biotech is no different.

Thus, the meagre levels of initial investment and the reluctance of public markets to make up for VCs failure to invest properly should result in small, unambitious companies, but those should nevertheless be capable of survival and, if reasonably lucky, grow to fill their more mundane markets. The critical skill is to make sure that money does not go out faster than it can be brought in, as investment or sales. This is the task of effective, commercially aware management.

So it is not surprising that VCs state that management is the single biggest factor in determining which ventures VC will invest in, and the research literature on how VC's work supports their statements [20, 67, 89, 109, 139–41]. Of course, different investors will look at different aspects of a business plan, and the business behind it, to reflect their interest. However, the majority of VCs investing in early- to mid-stage biotechnology companies have a remarkably consistent view of what is important to evaluate. After screening the business opportunity for the basic fit with their investment criteria, the most important factor stated by nearly all studies is 'management quality and experience'. Manigart [20] analysed VC appraisal of projects, and found their perception of the risk factors a proposal to identify management skill issues as top, then product, then management financial contribution, then the expected time to exit. Last on the list was the nature of capital market (compare this to the complaints about capital markets commented on in Chapter 6). MacMillan et al. [142] analysed what VCs thought were successful versus unsuccessful ventures. In a high-tech context VCs identify a protected product that can be brought to market and a management team capable of the sustained effort needed to bring it to market as the primary distinguishing factors. The major difference between the successful cases and otherwise similar companies was the management teams capacity for sustained effort. One VC commented that a good management team with poor science was preferable to a poor team with good science. Certainly true that strong business management is essential, no matter how strong the technical skills of the founders (see, for example, [143] on the story of Shockley Transistor versus Fairchild and Intel). However, Oakley [144] argues that such skills can be taught, and the entrepreneur turned into a manager. In line with this, VCs with a more 'hands on' management style will invest in weaker (i.e. less experienced) management teams because if the company is well funded it will attract strong management [138]. This is also in line with Rathman's comment on management noted above (Chapter 4) [87].

Riquelmo [89] summarises the requirements put on a new venture by VC's understanding of what makes a successful:

- Management team
  - relevant business experience,
  - have worked together before,
  - balance of skills,
  - personality conducive to business (includes leadership, teamwork, initiative, honesty, communication, confidence).
- Market and product
  - new product,
  - niche market,
  - growing market,
  - competitive advantage,

- ○ technically advanced,
  - ○ new use/unsatisfied need.
- Financial outcome

Interestingly, Riquelmo [89] also comments that high on a VC's requirements is a product that is at prototype stage (i.e. not research), and
whose development is not capital intensive, neither of which apply to
drug discovery, but which do apply to other applications of biotechnology. Despite this, companies engaging in drug discovery and development are preferred as investments by nearly all 'biotech' VCs, as we shall
see in Chapter 8. This is because, as we have discussed, the 'product' for
a biotechnology investment is *not* the product of the science, but is an
investment product. In that sense, the product (the company) is indeed
a prototype or nascent public company, and indeed will not require
unnervingly vast sums of capital to take to an exit if the public markets
are right.

Thus, the management team is universally stated to be of paramount
importance, with the actual science and technology of the company
usually coming second. Walton [145], a partner at Oxford Bioscience
Partners, comments that when he joined VC 'I was told by my partners
that they always backed "management, management, management"'.
We have reviewed the 'classic' literature statements of the importance
of management to VC above. This quote from a prominent UK VC is
typical: 'First you need a science degree or a PhD. . . . Second, you need
to secure sufficient funding. "Without money the science stops" says
Evans. . . . Third, and most important, you must set up a management
team. "You cannot concentrate on the science and do deals on the
side"' [146].

It is also believed that better founding management can compensate
for lower initial capital – it is possible to replace financial capital with
human capital [147]. Conversely, if you are planning to undercapitalise
the company (as we have seen European biotech VCs systematically do),
the very highest quality human capital is essential. So when looking at
an investment opportunity, both the literature and the mythology of
VC says that management is of primary importance, over even science,
patents and products.

This is an extraordinary proposition for three reasons. The obvious
one is that without scientific leadership a biotech company is little more
than an investment scam, and those who actually have made company
successes in the industry, such as Rathman (quoted above) recognise
this. Management follows excellence, it does not get there first.

The second is that VC actions are completely at variance with the proposition that they invest in strong management teams, as we shall discuss in a few pages. The third, however, is more closely related to the investment decision, and that is VCs do not spend nearly as much time or effort on evaluating the management team as their statements of its primacy would imply.

Mason [138] analysed how VCs *actually* evaluate early stage business plans sent to them, with results summarised in Figure 7.1.

Only 12% of the time is spent analysing what VCs say is the most important factor, the entrepreneur/management at the heart of the endeavor: nearly twice as much is spent on analysis of the market for the product (and this is a company that has not yet created its product, and so cannot estimate the market with any accuracy) and the company financials (for a company whose future operations are fundamentally uncertain).

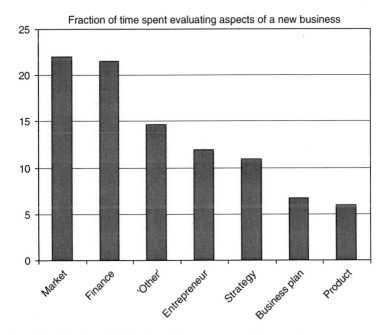

Figure 7.1 How long VCs look at different aspects of propositions

Note: Time spent on evaluating different aspects of a business plan. Verbal protocol analysis, that is, analysis of the actual fraction of time spent considering an issue in several VC groups as they evaluated incoming business plans.

Source: Drawn from data taken from [138].

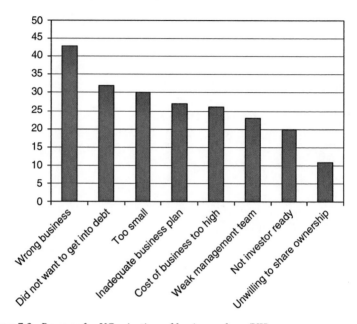

Figure 7.2   Reasons for VC rejection of business plans (UK)
Source: Data from [119], analysing data from Global Entrepreneurship Monitor (GEM).

Mason [139] also performed a verbal protocol analysis, which dissects the decision-making process and asks how often a specific topic is brought up in the process of making a decision. The entrepreneur and management score relatively higher on this, achieving 29% of the 'hits'. This still does not suggest that this is a principle or dominating factor.

Reasons for rejection also do not support the hypothesis that VCs take substantial cognizance of the management team. Figure 7.2 shows data from one study on why plans were turned down by VCs.

The 'wrong business' column is artifactually high – many agents who offer to assist companies raise finance do so by sending the business plan to everyone they can think of, regardless of whether it might meet the investor's criteria for investment. (I can attest to this from personal experience – in the late 1990s Merlin Ventures stated clearly that it invested only in biotech companies in the UK – despite this, around 40% of plans submitted for review were from companies that were not in biotech, not in the UK or both). However, even disregarding this, 'management' issues come low on the list, which either means that management teams are of very high quality in Europe (not an argument that has found favour, as we have discussed), or that the management team is not the primary focus of analysis.

Why do VCs apparently not care so much about the management of companies as their public statements would indicate? One reason is that they intend to take control of the management of the company themselves.

The classic structure of a UK company is that the company is controlled by the Board of Directors on behalf of the shareholders. French companies have a similar structure, German companies have a Supervisory Board with less direct powers. US companies have a similar model to the UK. However, the broad thrust is similar – there is a group of people whose role is to run the company for the benefit of the stakeholders, and with a legal responsibility to do so in an appropriate manner. To make operations practical, the Board of Directors (or equivalent – I will adhere to UK usage hereafter) appoint managers to execute aspects of the company's operations, and delegate powers to them. The Board of British Airways could decide that flying the planes was the most important aspect of the business and do it themselves – however, for a variety of practical reasons it chooses to hire pilots to fly the planes for it, and gives those pilots absolute discretion about most aspects of flying: it is impractical, for example, for a pilot to phone up corporate headquarters and get Board approval for an emergency landing. If Boards meet once a month (typical for a UK biotechnology company), matters that are likely to have to be resolved on a shorter timescale should be largely delegated. Other matters, such as technical decisions on projects, routine labwork and other minor operational matters should be delegated: indeed the CEO should not concern himself with them. The managers report to the Board. Usually the most senior managers are also Board members. This is what Company Governance 101 suggests. The Board then reports to the investors.

The reality in VC-backed entrepreneurial companies is completely different from this model. In a VC-backed company, the management is controlled completely by the VCs: in effect, VCs manage the company in all important aspects. This is accomplished through two mechanisms: Board control and shareholder veto powers. The effect is that the entrepreneurial management of the companies is relegated to being the junior executive level of the *real* management, the investors. Again, this is in such variance with the perceived state of affairs that it is worthwhile expanding on how this situation arises before discussing the effects on the company.

When VCs invest they will usually insist on Board representation. However, all VC shareholders will also require the company, and its Board

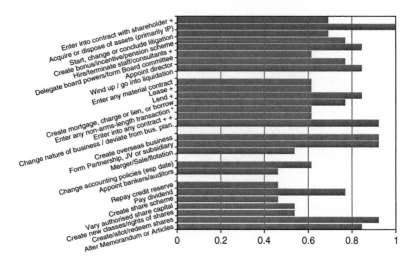

Figure 7.3   Shareholder veto powers
Note: Collected from the author's analysis of 13 investments in UK biotechnology companies 2001–5, involving 30 different investment management groups.
Source: Figure redrawn from [148] with kind permission of Journal of Commercial Law Studies.

and managers, to agree to terms and conditions embodied in a contract called the Shareholders' Agreement (Founding Shareholders' Agreement if it is at start-up). This specifies a wide range of acts that the shareholders agree to as a condition of receiving the investment. As this is a private contract between shareholders it is not part of the public documents that are a matter of record in the UK or in many other countries. However, the Shareholders Agreement effectively abolishes the ability of the Board and the management it appoints to act independently of the investors.

I have reviewed the terms of the 13 Shareholder Agreements covering investments in UK biotechnology companies and involving 30 different specialist investment groups. Figure 7.3 shows the powers required by the investors, and the fraction of those agreements that demanded them. (Because these documents are considered commercially sensitive and very confidential, the sample has an unrepresentatively large number of Cambridge-based companies in it, as they are the companies I was able to access for this data: however, my discussions with those involved in company financing elsewhere in the country suggests that they are not atypical.)

For the shareholders to require control over liquidation or sale of the company, creation of new classes of shares and formation of subsidiaries is reasonable. Were they not to have such control, the management could form a subsidiary which the management owned, pass all the company's valuable assets to it, and leave the investors with a valueless shell. (The investors would then sue them, of course, as such an action would be illegal as it is in clear breach of the Director's fiduciary duties to protect shareholder interests, but going to court is expensive and time-consuming, and who wants the hassle when the action can be blocked at the start?)

But typical terms also give shareholders (not just Directors) control over appointment of staff, signing of contracts and leases, acquisition or disposal of IP and 'any material change to the business'. As biotech companies change constantly, this is equivalent to saying that anything the CEO wants to do, from hiring a new PA to selling off a patent that is not core to the business to changing the direction of the research plans requires sign off from the VC. This is not just information. This requires formal, often written approval.

The veto powers also give VC shareholders control over the appointment of Directors. So as well as requiring board representation, they have control over any Board appointment. New Board appointments are therefore in the sole gift of investors, and as a consequence the investors effectively control all the other members of the Board through having the power to appoint other Board members to outvote the uncooperative or obstreperous.

Added into this is requirement for provision of information more appropriate to middle layers of management than to the Board. The Shareholders' agreement contract usually provides for the VC shareholders (again, not just the Board members) to receive detailed financial reports once a month. In addition, it is common for Boards to require detailed technical reports on individual projects, sometimes even individual experiments, to require complete lists of all business development contacts and the state of their progression. These information requirements are at least as extensive as the veto powers illustrated in Figure 7.3.

The net effect is that the VC acts as a layer of control over the week-to-week operations of the company, quite separate from their duties as a director, and the control they can exert simply by not giving the company the cash it needs. Fried [149] also comments on this unique control structure in companies in the US: indeed these practices were invented in the US and copied in Europe, as were a number of other types of terms imposed on companies by European VCs in the mid 1990s onwards [150].

In line with this extraordinary level of contractual control on the Board, management theory states that VCs do not just provide finance to companies, but also provide a range of management and strategic support. It is stated that a major difference between a venture capitalist and a bank is that the former supports their investee entrepreneurs in non-financial ways as well as with money [12, 71]. This additional support is considered a valuable part of involving a VC with your company. The entrepreneur, it is argued, is likely to have much less commercial experience than the VC, and this limited business experience means that they benefit from the business experience and advice of the VC. Support can include expertise, networking, contacts, sounding board for ideas, help developing strategy and identifying and recruiting top managers [12, 151–2]. VCs also have commercial contacts, leverage with potential clients and suppliers and a range of other value-adding features which mean that their participation in the *management* of the company is valuable as well as their participation in its *direction*. Or so the theory goes.

This 'help' is not always appreciated. Some entrepreneurs believe they do not need the help – they will tell the VC 'I don't need your help; I just need your money' [153]. Others may feel that VC 'help' is just 'interference' in their new venture. However, it is a widely promulgated view that, wanted or not, VC brings a range of non-monetary value to investee companies, to supplement entrepreneurs' limited business experience, networks and capabilities [12, 151].

This, in any case, is the US story. Rothwell [154] pointed out that in the early days of VC in Europe 'VC' did not behave like VC in the USA. It provided negligible non-financial support, and was essentially a high-risk, high-return institutional investment mechanism. Since then European VC has claimed that it does provide substantial support for the companies, mimicking the American model. The following quote from Soffinova's European website (2006) is typical:

> We are constructive investors, assisting our portfolio companies with much more than money. Our mission is to create value by providing talented entrepreneurs with the resources, experience and network necessary to turn their bright ideas into great businesses.

This added support is a major selling point for VCs. European VCs now widely promise that, as well as money, they will deliver support to the companies they invest in. An analysis of some of the support offered by European biotech VCs is given in Figure 7.4.

Thus, both the legal structures imposed on a start-up company and the management literature and VC PR material strongly suggest that

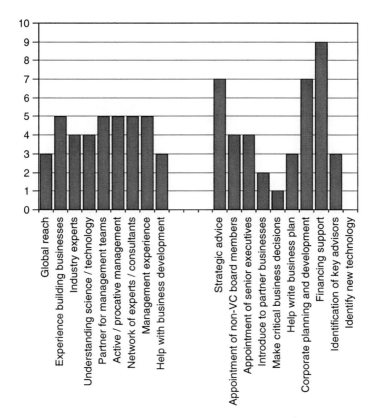

Figure 7.4   European VC claims for non-financial support for companies
Note: What European VC investors in biotech say that they do for their companies other than provide funds. Activities are grouped into 'strategic' and 'tactical', but in reality there is a lot of overlap between these classes.
Source: Data from 13 VC websites and presentations, summer 2006 [21].

VCs take an active role in the management of early stage biotechnology companies in Europe. But does this actually happen? Two measures of outcome suggest that it does not.

Firstly, we can ask whether VCs actually help European companies achieve financial outcomes. Manigart [155] followed 565 Belgium VC-backed and 565 non-backed companies, and found no difference in survival: the 'VC value add' did not seem to make any difference. Busenitz [156] followed a portfolio of VC-backed companies for ten years, following what happened to them as a function of what the VCs did. VCs were categorised by their own statements into companies that provided intensive advice and guidance and ones that were 'passive investors'. The study found that VCs providing advice and information on strategic matters did not affect outcome at all. They did, however, find that ventures treated in a 'procedurally fair' manner had improved outcomes (in terms of financial reward to all concerned): this means avoiding some of the VC practices tagged 'predatory'. MacMillan [157] categorises VCs into 'Laissez Faire', 'Moderate Involvement' and 'Close Tracker'. The investments made by investors categorised as 'Laissez Faire' do better (by the VC company's own judgement) than ones which have more VC involvement. MacMillan finds no systematic factor in which companies VCs chose to follow closely: The level of involvement is not dictated by economic theory or by practical experience: it seems to be entirely matter of taste for individual VC.

Indeed, an analysis of advice of any sort provided to start-up companies suggests that the various categories of advice provided by VC is actually not all that useful. Aldrich [158] studies the cost and outcome of different types of advice to start-up companies (not just biotech), and found that although 'financial' and 'legal' advice cost the most, 'expert' advice (i.e. advice based on expert knowledge of products or markets) was the most valuable in terms of business processes improved (Figure 7.5).

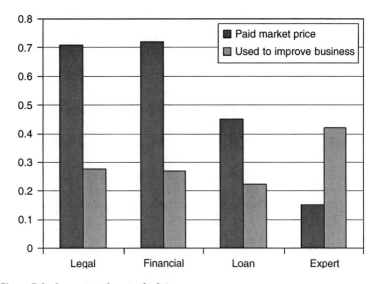

Figure 7.5   Impact and cost of advice
Note: Fraction of companies that had received advice on financial, legal, interim/loan financing and 'expert' areas of their business who paid full market price for the advice and who found the advice was useful for improving their business processes.
Source: Data from [158]: all companies analysed are UK SMEs.

While good financial systems put in place by a professional are obviously essential, the 'strategic' advice they give appears from this to be overvalued. This is the dominant type of advice promoted by VCs as being the most valuable. O'Regan [159] confirms that the most important aspect of advice, networking and training is company's awareness of developments in their technologies and markets, which they maintain through trade journals, networks groups, customer (=collaborator) contact, trade fairs etc. 'Strategic' and financial advice, whether provided by formal consultants or other experts, is rarely useful.

Secondly, regardless of whether the support is actually any use, we can also ask whether VC claims are internally consistent. Is it possible for VCs to provide the support they claim?

Studies of US VC companies suggest that 'Venture capitalists spend approximately one half of their time monitoring an average of nine funded ventures, which requires about 110 hours per firm per year' [160]. Zider [161] suggests that about 2–3 of a VC's time is spent this way based on US examples. Gorman [121] reports that US VCs in the 1970s and 1980s visited 'their' companies an average of 19 times/year, ~, about 100 hours contact time per company (this does not include travel time, desk work 'back at the office' and so on, which would at least double this) [71].

However, my observations over the period 1996 to 2002 suggests that 20% is a more typical figure for UK VC (Figure 7.6). This UK figure is more in line with the emphasis of the literature on VC, which shows that by far the majority of their business success comes from attracting funds, investing well and setting up controls on the investee companies as part of the investment process, and relatively little from managing the companies after investment [71].

We can quantify this in three ways: the amount of time and effort needed to raise a new venture fund, the time spent investing it and how many companies the remaining time is divided among.

Formal analysis of how long it takes to raise a VC fund has not been done. However, it is a process akin to a company seeking investment, which is widely estimated to take the entire senior management team around six months working time, very comparable to the time taken for a company to get investment from a VC, and for similar reasons (c.f. Figure 5.5). If VC management groups raise a new fund every three years (a typical average for European biotech in the period 1995–2005 as we shall see in Chapter 11), this means one-sixth of their time on average will be spent raising funds.

In addition, the average time taken to invest an amount of £10 million or less in a biotechnology company is about 320 days (Figure 5.5). About

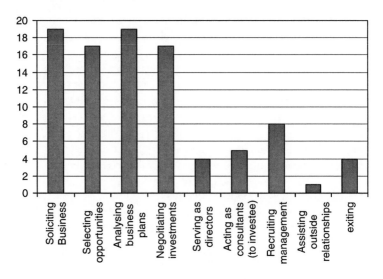

Figure 7.6   Fraction of VC time spent in different activities
Source: Data from my informal observations. The activity categories are the same as those used by Zider [161].

half of this is the time taken to negotiate with VC groups on all aspects of the investment (a process which typically generates a stack of definitive documents – called 'The Bible' – several inches thick for each investment). Thus, for each investment, a VC management team can be expected to spend a significant fraction of 160 days per investment on 'due diligence' and the subsequent negotiation. This is distinct from legal, intellectual property and financial 'due diligence', which can be outsourced.

An average active VC house makes between 0.5 and 0.7 investments in new projects (i.e. ones in which they are not following a previous investment) per senior staff member per year (Figure 7.7).

Thus, each senior figure can be expected to spend between a quarter and a third of their time on making the investments that they successfully conclude. As investment houses reject the vast majority of the proposals put to them [21], we can at least assign the same amount of time to evaluating the investments that are not made: even if this is done fast and efficiently, the criteria used require effort from senior staff [20, 138]. This implies that an energetic VC who works 70 hours/week spends the majority of their time just in the process of investing, without considering the time taken for raising funds and the complex and time-consuming process of achieving exits from investments.

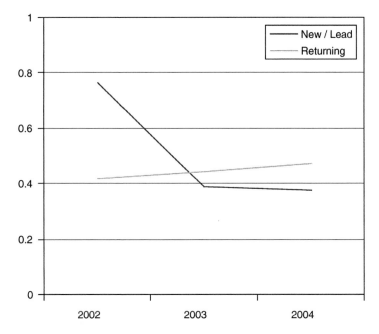

Figure 7.7 Investments per partner per year

Note: Number of new and 'returning' investments per senior staff member per year. Investment numbers from my own databases. Staff members from VC websites for the most active 12 VC groups investing in biotech in this period. All staff involved in the investment process are included in the numbers, not just 'partners'.

This and fund-raising therefore accounts for between 70% and 83% of a VC's time. The rest is spent supporting investee companies, including achieving exits for their investment, or between 50 and 90 days/year (based on a 300-day working year, which is not implausible – I set aside claims that VCs work over 100 hours a week [162] as being implausible PR).

Typically, investor directors sit on the boards of several companies. VC literature and websites usually list the principles as sitting on two or three boards. However, this is not a true representation of their effort. A search of the websites of 12 VCs and the websites of their investee companies reveals that those that sit on any boards at all (i.e. are considered sufficiently senior to have an investee company board seat) often sit on more than this: an average of seven or eight is typical (Figure 7.8). But this again is voluntarily supplied information. The only complete analysis of Board seats is from information that

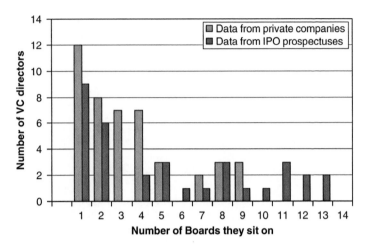

Figure 7.8   How many Boards do VC investors sit on?
Source: Data from IPO prospectuses kindly provided by 17 public European biotech companies. Private data from websites of a further 38 companies.

has been legally verified, and a good source of this is prospectuses companies produce for IPO. Figure 7.8 shows the number of board places in active, investee companies (i.e. excluding shell or holding companies and excluding the companies of the investment group itself) held by investor directors of European biotechnology companies at the time of their flotation. Note that these are still private companies at this time: this is the prospectus stating the structure of the company *before* IPO, so these VCs are in effect acting as a layer of management as discussed.

It is not clear that even this accurately states the extent of investor directors' active involvement with other companies. For example, the ReNeuron IPO prospectus of November 2000 stated that the investor director had a 'previous' (i.e. currently inactive) directorship in Amedis Pharmaceuticals (p. 86). However, company records and Amedis Pharmaceuticals' website at the time listed this director as active, and as the company was raising finance at that time it is unlikely that this took up none of this director's attention, so this was not a 'previous directorship' but an active one.

This shows that a 'typical' investor director in Europe – principally the UK – sits on a large number of company boards. From a company perspective, you are more likely to get one of the 'prolific' directors on your board than one of the 'scarce' ones: analysis shows that you have a 50% chance of sharing 'your' investor director with at least nine other active companies. He or she will be required (in theory) to turn up to monthly board meetings for each of these companies, having read any board papers beforehand, which will take up around a day each. For ten companies, this is 120 days/year. Even 12 visits for formal Board meetings exceeds the amount of time the VC has left after raising money, making investments and looking for exits. There is no time left for any additional support at all: indeed it would not be surprising if the anecdotal stories of VC directors consistently missing board meetings unless a crisis has already arisen in the company are bourn out as statistical generality; in Europe, they simply do not have time.

Entrepreneurs are not the only ones to find this disturbing. In the UK and the US the regulatory climate is turning against multiple directorships, as a series of public accounting scandals have shown that the multiply-boarded non-executive directors (NEDs) of some companies had neither the time nor the skills to supervise their management properly. The recruitment agency Korn Ferry was finding, in the late 1990s, that many directors believe that too many directorships prevent proper execution of their directorial duties, and that the most common reason for declining directorships is that the candidate will not have enough time to do the job [163]. The Council of Institutional Investors suggests that a top executive should not have more than two NED seats as well as a full-time job: the National Association of Corporate Directors suggests three to four NEDs plus a full-time job. Ferris [164], looking at major corporations with over $100 million assets, finds that having multiple directorships is not a problem: they find no correlation between number of directorships held and company performance, committee membership or securities fraud actions, that is, no failure or reduction in oversight activities. So the issue is not failure of oversight from multiple NED seats, it is failure of 'help'.

The situation is very different for business angels. Their model is to invest their own money (we will come to the critical importance of this), and hence they have a strong motive for increasing the value of their investment in any way they can without investing more funds. Mason [138] showed that whereas VCs minimise risk by screening out risk, business angels minimise it by management post-investment. This is why business angels will often say that their reason for not investing in a new opportunity is not that they do not have the cash (although this is often the case) but because they do not have the *time*.

At the start of this chapter I mentioned three reasons for disbelieving VC claims about the pre-eminence of management among the criteria for a good company. The clear importance of the science and technology in the business case and in VC decision making is the first. VC management takeover (without concomitant VC time input) was a second. The third, though, is their treatment of the management that they state they have selected with such care. VCs do not behave towards management as if they were the most valuable assets that the company has invested in. I will expand on this here as it properly lies in the discussion of how VCs actually *manage* companies in European biotech, and do it badly.

The most obvious expression of this is that the first thing a VC does when a company 'goes bad' is fire the management. Ruhnka et al studied this phenomenon at length, and found it true regardless of whether the problem was the management, or management-related and regardless of whether firing them has any effect on the problem [165]. This study found that changing the management has almost no correlation with return on investment – i.e. it was not productive even in the VC's own frame of reference. This is particularly true of inexperienced VC teams [166]: VCs with extensive experience tend to respond to problems (real or perceived) with constructive and active approaches rather than the negative one of demanding a severance between the VC and the entrepreneurial management.

The tendency to remove the management from any involvement in the company is particularly notable for the originating venture team, which is routinely thrown out of companies when VC investors come in, regardless of whether the company is 'going bad' or not. Indeed, it is almost inevitable that such New Venture Team Members (NVTMs) replacement is not a result of company failure, as these are (by definition) new companies that are sufficiently promising to receive VC investment. NVTM replacement seems to be a general rule: one European VC was quoted as saying

> It is certainly not inevitable that we have to change one of more members of the original management team. We have done 75 deals and we only made changes in 73 of those deals.
>
> [167]

Removing the founding management from any significant role in company management or direction, recruiting a highly paid CEO to replace them and then firing the CEO in turn is apparently standard practice for European biotech VC.

This approach is not confined to Europe: Hellman [168], examining silicon valley companies, found that VCs usually played a central role in replacing founder with an outside CEO, and that firms are more than twice as likely to have a CEO turnover event (i.e. change of CEO) once they have VC. However, unlike in the UK, out of 38 of these 'turnover events', the founder remained associated with the company (as an executive or a director) in 28 of them, which was not much different from the ratio in non-VC companies that changed management. In Europe, the fraction of companies retaining their founder in any meaningful role after a VC-induced 'turnover event' would be substantially lower.

In European venture-backed biotechnology companies complete replacement of founder management and directorial roles is the usual route to management succession [22]: 'management succession' is code for 'fire the founder', in agreement with Boeker and Karichalil's [169] finding that founder replacement is more frequent in firms with low founder ownership and fast growth. Some VC investors are particularly adamant that the founder be removed from a management role in the company quickly to make way for 'professional' management.

Typically, the VC will bring in a new CEO of their selection fairly soon after investment. Formally, this first incoming CEO is there to stabilise the company post-start-up, strip out some of the founder's less commercial enthusiasms, and build up a management team. Informally, they are also there to act as the VC's proxy for the 'fire the founder' moment. The survival of the NVTMs in UK biotechnology companies is shown in Figure 7.9. Academic founders remain associated with the company for the longest time, but usually only in roles such as members of the Scientific Advisory Board, which has no legal or executive power. They are 'kicked upstairs' to this role after an average of slightly over two years

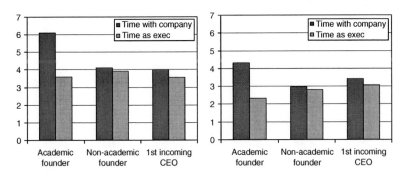

Figure 7.9   Survival of NVT after foundation
Note: Average times NVTMs stay with as VC-funded start-up company, after company foundation (left panel) and after receiving first VC funding (right panel). Shown are times in an executive role, and times in any relationship with the company. Data from my own database on all biotech companies funded by the major biotech VC investors between 1998 and 2003. Left panel – NVMT survival after foundation, right panel – after first VC funding.
Source: Reproduced from [170] with kind permission of Palgrave Macmillan.

in a meaningful role. The first CEO of the company and non-academic founders are usually not retained substantially after their executive role ceases, which happens around three years after investment.

This agrees with Walton's observation [145] that the average biotech company gets through 3.5 CEOs between foundation and IPO (a time of five years in the US sample Walton was examining).

If investment is based as much on the team as on the science, why fire key members of it an average of 2.5 years after investment? The universal answer is to bring 'professional management' into the company. The qualities that lead an entrepreneur to create a new business are probably different from those that are needed to run that business. So management literature holds that there comes a time in the growth of any company when the founder-manager must relinquish control over some key aspects of the company's function if the company is to flourish [171–4]. Often they are loath to do so: this 'succession crisis' is a feature found in several empirical studies of the evolution of new, growing companies [175–6]. In line with theory, some studies show that companies that fail to manage management succession effectively perform less well (see, for example, [143, 177–80]. In high-tech, high-growth companies the need for management change may be particularly acute [181], although Willard and Kreuger [182] find that replacing the founder-managers by 'professional' management did not affect the productivity of high-tech manufacturing companies. In reality, in any case (as we will see in more detail in Chapter 10) the management brought in is often mediocre and expensive, and sometimes catastrophically poor.

VCs play a dominant role in 'management succession' in VC-backed companies [168]. Clearly, firing some or all of the NVTMs does not benefit the NVTM, personally, psychologically or financially. Because founders (usually) do not want to be fired from 'their' company, this brings the VC and the founder team into conflict, a conflict which the VC regards as constructive at a low level but destructive if (as is almost inevitable) it becomes intense or 'personal' [183]. This conflict is the source of a lot of the angst and vituperation that echoes round the Internet regarding how terrible VCs are. But for our purpose we should ask not whether firing the founder is bad for the founder, but whether it is good for the company. The arguments above say that it is for the benefit of the company (and hence the shareholders). However, the data strongly suggests that it is not.

We can analyse this by looking at the exit that the VCs themselves want – an IPO – and asking if firing the NVT helps towards this. This is done indirectly in Table 7.1.

Table 7.1 NVT stay and company success

	Number of companies	VC money raised	Number of employees today (or at peak)	Prods in clinic	Exit
All out in 4 years	10	11.75 (10.35–13.15)	19.5 (15.43–23.57)	0.5 (0.38–0.62)	0.0 (0.0–0.0)
Mixed	23	14.62 (13.51–15.74)	16.39 (13.86–18.93)	1.0 (0.86–1.14)	0.26 (0.2–0.32)
All in at 4 years	44	12.48 (11.58–13.38)	24.36 (20.89–27.83)	4.49 (4.23–4.75)	0.30 (0.27–0.33)

Note: Quantitative success outcomes by companies grouped according to whether all the founding team had left by the fourth anniversary of foundation ('All out in 4 years'), all the founding team were still with the company after four years ('All in at 4 years'), or whether some were with the company and some had left ('Mixed'). Values are averages for the companies concerned, with two-sigma ranges in bracket.

Source: Reproduced from [170] with kind permission of Palgrave Macmillan.

In terms of company size (as measured by the number of employees), the number of products in clinical development, or the fraction of companies that have achieved exits, companies that retained their NVTs did better than companies that did not. Only in terms of VC cash raised was there no significant difference between the classes, and as VC dogma says that firing the founders is an inherently good thing to do, it is not surprising that companies with poorer prospects in terms of growth and IPO can nevertheless attract equal VC funding through changing the management team fast and frequently.

Florin [184] found a similar result, looking at the returns on VC-backed and non-VC-backed companies two years after their IPO in 1996. He found that VC backing pre-IPO meant larger and more prestigious management teams, more capital value in company, and more chance that the founder was fired. However, post-IPO performance of VC-backed and non-VC backed companies are the same.

> Thus, results show that returns to founders decrease with VC involvement, with no significant improvement in terms of firm and investor performance after the IPO.

Busenitz [156] also found that firing a founding team member was negatively correlated with company success in a longitudinal study of 120 start-ups over ten years, and firing two or more was more strongly negatively correlated. The strongest *positive* correlation was with the size of the VC fund, and by implication the amount they could invest. Again, firing the NVT is a poor idea, investing adequately is a good one. A range of other studies has shown that management change is actually detrimental to project performance, and can negate the performance enhancement that the change is meant to accomplish (see e.g. [185] and refs therein). Florin [184] also found that VC involvement in companies resulted in reducing the chances that the founders remained in charge, but that this change did not improve performance of the company after IPO (i.e. when independent, third party investors could judge the value of the company).

Is this because good companies created by good NVTs get to IPO, whereas poor ones do not, and also require NVT replacement? Probably not. Figure 7.10 shows the average time that founders remain with companies compared to the success of those companies after IPO. After IPO companies can do well or badly, and their stock price relative to an industry sector index is a measure of this. Figure 7.10 shows that the sector-adjusted stock success of companies after IPO is not related to

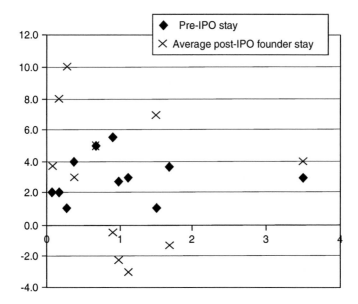

Figure 7.10   X-axis – share value today as a fraction of share value at IPO

Note: Post-IPO share performance versus NVTM stay. X axis – share value today versus share value at IPO, corrected for changes in HASDAQ composite over same period. Y axis: average time the NVTM stayed in the company: diamonds – NVT stay from foundation, crosses – NVT stay from IPO (note that some of these are negative as the NVT was dismissed before the IPO). Companies analysed are Alizyme, Ardana Bioscience, Ark Therapeutics, Biofocus, Cytomyx, Evolutec, Henderson Moreley, Oxford Biomedica, Pharmagene, ReNeuron, Vectura. The twelfth floated company – Fluid Technologies plc – did not have sufficient founder information to analyse.

Source: Reproduced from [170] with kind permission of Palgrave Macmillan.

how long the founders stay with the companies before IPO (Spearman's correlation = 0.08), but in line with the theory post-IPO stock performance was weakly and negatively correlated with how long the NVTMs stayed after IPO (Spearman's correlation = −0.22). So the argument that not firing the team is an effect of having a good company (created by a good team) does not hold, as the team would still be good after IPO.

In summary, keeping a founder associated with the company before IPO is positively correlated with performance before IPO, keeping them associated with a company before IPO is unrelated to performance after IPO, and keeping them associated after IPO is negatively associated with performance after IPO. So we can conclude that rapid removal of the founders, *as a standard rule of operations*, is not in the interest of shareholders who wish to realise their investment through an IPO. There are cases where the founder entrepreneur can be almost pathologically 'toxic' [153], creating damaging barriers between staff and management and management and VCs in order to preserve their position, and hiding or even blocking technical progress to retain control of the company: removing them in such cases is essential. But to use these rare cases as a reason to automatically get rid of the founding team is unjustified by the outcome that is achieved. As in another aspect of VC control of start-up biotech companies, their actions are exactly the opposite of those which would encourage company growth and success.

8
VC Effects on Business Efficacy

The last chapter described how VCs abrogate management authority, fail to provide the support they promise to management and instigate management change when it is counterproductive to the VC's own goals of stock value appreciation and IPO. An explanation for this rather odd set of observations is that the VC has a deep knowledge of how a biotech business should be run and, while they do not have the time to provide that understanding to the company management, they nevertheless can direct the management to that end, and simply get rid of them if they disagree rather than try to work with them to change their mode of operation. This could be effective if the investors themselves had a clear idea about, or a good track record in, managing the affairs of investee companies. This chapter tests this by examining the effects of investors on the company's business model, the next by examining their effects on the execution of that model.

The effect of VCs on the business model of their investee company is widely known, but little discussed. VCs have some very strong views on the business model that they want companies to pursue, arguing that these are the only models that 'can make money'.

There is a wide range of business that is captured by the idea of 'biotechnology'. As I mentioned briefly at the start, some do R&D, some sell products: among the second group some of their clients are other biotech companies, some are drug companies, a few brave souls sell to farmers or builders and many sell to the state-funded biomedical research industry. Analysis of the public companies (Figure 8.1) and private companies who seek investment (and hence presumably have ambitions to become public companies) (Figure 8.2) show this to be the case. Analysis of the totality of research-based companies calling themselves 'biotech' in the 1980s, that is, before VC decided to impose its idea of what

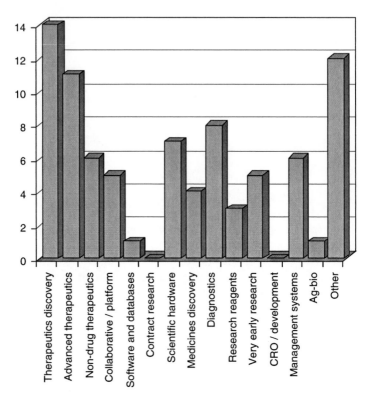

Figure 8.1   Business sector of public biotech companies (2004)
Note: Business model of public 'biotech' companies listed on London stock markets in 2004.
Source: Data from the author.

a biotechnology company was on the industry, shows a similar diversity of business goals (Figure 8.3).

The largest sector in 1988 was working on agricultural/environmental products (there is substantial overlap between the two in, for example, technologies for disposal of compost waste), an area which has not decreased in importance since 1988. Pharmaceuticals – either as finished drug products or as intermediates and technologies – are the subject of only one-eighth of the companies. And yet, the large majority of VC investments in biotech are in companies making new medicines. This is for reasons related to the business model of the VC, as I will expound later.

This has three important consequences. Firstly, companies that are not in the medicines creation business are not funded. The economic answer to this is 'tough' – many things are not funded by investment,

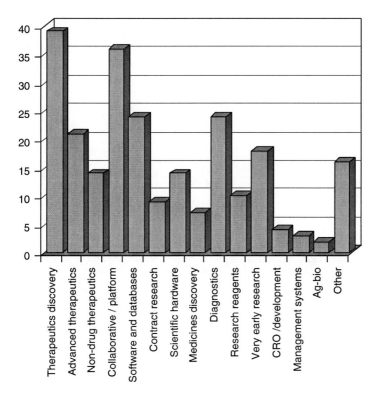

Figure 8.2 Business sector of private biotech companies (2004)
Note: Business sector being addressed by 83 proposals for investment which were considered for substantial due diligence by Merlin Biosciences 1998–2000.
Source: Data from author.

from trainspotting to moon landings. They either find other sources of funds (including sales) or they do not develop as commercial exercises. There is no argument for VCs funding non-drug companies just because they are there, any more than there is for funding the battalions of state-sponsored 'start-ups' that Europe generated in the 1990s.

Secondly, it encourages people to create medicines development companies, and not direct their research, energy and creativity to other ends. This is not really VC's fault, but it is an unintended consequence of VC investing patterns, which we will discuss in Chapter 12.

Lastly, though it is a very poor business model. For companies that deliberately chose this business model, this is the rod they have made for their own back. For others, though, they can be driven towards the

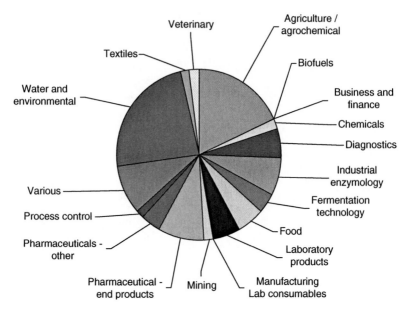

Figure 8.3   Industry areas of research-based biotech companies in 1988
Note: Data from the same company set as Figure 4.1. 'Research' company is here defined as any company with a significant amount of in-house R&D aimed at its own products – contract research organisations or consultancies are not included.
Source: Data from [79].

medicines development model rather than other, more sustainable business models, and suffer as a consequence.

As so much of the PR around biotech is about health care, it is worthwhile digressing for a few pages to discuss what a 'medicines development' company is, and why it is a poor business.[1] Those of my readers who wish to skip this slightly technical section, can go to page 114.

The rather clumsy phrase I have adopted here – medicines development – reflects a reality that most in biotech would prefer to forget. People do not take drugs (apart from in informal and usually illegal circumstances). They use medicines. A medicine contains a drug, other materials to turn it into a tablet or a syrup or a syringe full of liquid and comes with associated packaging, prescription advice and social acceptance in the form of the stamp of regulatory approval. It is a product for a specific disease, whereas a drug is a chemical that interacts with a specific chemical system of the body. Standard 350mg Aspirin tablets are medicines for headaches; 80mg aspirin tablets are medicines for reducing the risk of repeat heart attacks. The *drug* is the same, the *medicine* is

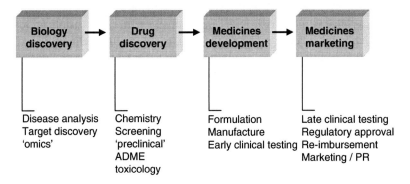

| Biology discovery | → | Drug discovery | → | Medicines development | → | Medicines marketing |

Disease analysis	Chemistry	Formulation	Late clinical testing
Target discovery	Screening	Manufacture	Regulatory approval
'omics'	'preclinical'	Early clinical testing	Re-imbursement
	ADME		Marketing / PR
	toxicology		

Figure 8.4   The medicines development process

different. This is a critical difference, because it is *medicines* that are sold to ill people and their doctors, not *drugs*.

The process of getting you or your hospital to buy a new medicine can therefore be divided into four phases (Figure 8.4).

The cognoscenti will recognise overlap and recursion between all of these – these are complications that I have addressed elsewhere [10]. 'Biology discovery' is finding how a disease happens, and what might be done about it. When a newspaper announces that 'scientists' have found 'the gene' for something and that a cure is only years away, this is the box those scientists are operating in. 'Drug discovery' is finding a chemical that affects that disease process, and which is also suitable to be put into patients (for example, chemicals which poison one part of the body faster than they cure another part are not good candidates). When 'scientists' have 'found a new wonder drug' for a disease, this is the box they are working in. Then that drug has to be turned into at least a prototype medicine, tested that it is safe in animals and people in that order and tested to see if it has some effect. This is 'medicines development'. After that the pharmaceutical company has to convince us that it is a good idea to use the drug – this it does by more extensive clinical tests, by getting approval from the appropriate regulatory authority to sell it as a treatment for a disease and by extensive marketing and PR campaigns. (From a societal point of view, regulatory agencies like the Food and Drug Administration (FDA) and the European Medicines Agency (EMEA) might be there to improve health care, but as far as the pharmaceutical industry is concerned they are there to help them market a drug by providing authoritative verification of the Drug Label – the words on the box that say 'can be used to treat disease

X' – which allow the pharmaceutical company to charge high prices). For the consumer, pharmaceutical marketing is the adverts they see for brand name medicines, but this is just the tip of a PR iceberg that includes scientific papers and conferences about the *drug* as well as the regulatory process, thick reports and dossiers of information, post-launch clinical trials and other material on the *medicine.*

The whole process takes an average of 12 years, has a 95% chance of failing for any one project, and costs $750 million (including the costs of failures) to create one genuinely new drug [10]. As a result, most new medicines launched do not contain new drugs – they take old drugs and start at the 'medicines development' stage, finding a new use or a new type of tablet for them. This is still risky, but not as bad as starting with a business plan that says, 'Let's find out how cancer works and cure it!'.

I have analysed this process in a lot more detail than is described here [10]. In summary, the reason that the process is slow, risky and expensive is not due to any specific technical or procedural failing. The whole process is just hard to do.

Biotech companies working in the health care space therefore have three options for making money out of the science, technology and team that they believe can help to do this better than anyone else. They can sell technology or products for doing part of the process, do part of the process themselves or do the whole process. I will examine the business rationale for these three, in reverse order.

Doing the whole process involves setting up what is known as Fully Integrated Pharmaceutical Company (FIPCO), with everything from basic biological research to sales and distribution. If we are to pursue a scientific or technological business advantage, then it does not make sense to set up a FIPCO doing what all the existing pharmaceutical companies are doing, so a brand-new FIPCO should target really radical therapeutic modalities. This is what all the first wave of huge success stories in biotechnology did – Genentech, Amgen, Genzyme, Biogen, Serono, Celltech – and some of the spectacular failures such as Centocor (whose principle product Centoxin bombed in clinical trials, wiping out most of the company's value [186–7] – notably, the company survived this and still flourishes today). These companies' technology – recombinant DNA to make protein drugs – was at the time radical cutting edge of science, never mind commercialisation, equivalent to dot.com in 1994 or nanotechnology today. Because, they were doing the science in a radically different way from anyone else, they had the chance to do it far better: because they were very well funded, very bright and (we must admit) quite lucky, they were enormously successful. This is therefore a model that can work well, in the right time and place.

Despite this model for success, the FIPCO model has fallen from favour, especially in Europe [188–9], because of the enormous investment needed: Stoiber comments that 'Product companies are crushed between high financing needs and their cost of capital. A full integration strategy can destroy value for founders and early investors' [190] (although it made a brief comeback in 2000 when VCs saw a very strong bull public market to support their investment model [191]). Genentech absorbed about $1 billion in investment before it broke even, and although it is now capitalised at over $80 billion [32] it is believed that $1 billion is more than any plausible investment mechanism can provide (although recall in Chapter 6 that we saw that the US genomics companies did raise these sorts of funds when the market was ready for them, so the reality is that, with brilliance, work and luck you *could* do another Genentech today).

Today's radical European technology companies in stem cells or gene therapy are small private companies or, in the case of the few public stem cell companies such as Stem Cell Sciences and ReNeuron, small-cap public companies which have received less than 1% of what Amgen and Genentech received to make them global giants. Of importance here, we can note that this is not a failure of public markets to support the companies. As examples, we can compare ReNeuron (a leading and technologically adventurous stem cell company founded in the UK in 1998) with Geron (a leading and technologically adventurous stem cell company founded in California in 1992). ReNeuron received £5 million before it floated in 2001, compared to the $40 million (£26 million at the prevailing exchange rates) put into Geron before its IPO in July 1996: in accordance with Figure 6.1 Geron was valued pre-money at IPO at $64 million (~£41 million at the prevailing exchange rate, or about £46 million in 2001 pounds), ReNeuron at around £20 million, a price that the markets immediately decided was too high and punished by reducing very substantially. The markets recognised the value of promising technology backed by solid investment. Comparison with other disruptive technology companies such as Google suggest that when public markets are presented with an outstanding opportunity to get ahead of the curve they can usually recognise it.

The Geron/ReNeuron comparison emphasises again the role of *private* investment in supporting companies in the long term. ReNeuron's share price plummeted after its IPO in late 2000: neither its VC backers nor the public markets supported it with further investment, and the share price

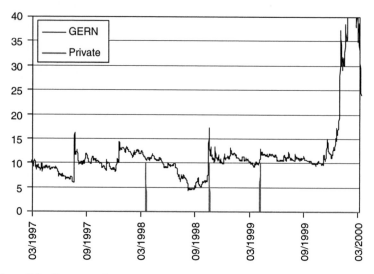

Figure 8.5   Geron stock price
Note: Geron stock price 1997–2000. Red bars are private investment rounds announced by the company.
Source: Data from Geron website, NASDAQ and CapitalIQ.com.

fell to ~5% of its IPO value before it was taken off the public markets. Geron's share price also fell over several periods, notably in 1997–8, a fall arrested by private investment which both supported the company and demonstrated confidence and commitment (see Figure 8.5). Other upticks in stock price, in July and December 1997 relate to patent and collaboration successes respectively, again illustrating the value of these events. This is completely the reverse of the desire expressed by a UK VC that companies should 'cut loose' their VC support as soon as they could after IPO [130].

If you cannot do the whole process of medicines discovery, development and sales as a FIPCO, then you can seek to do the bit that you believe you are good at, acquiring intellectual property from companies or research institutes 'upstream' and licensing the IP with value added to those companies downstream. This offloads the enormous timescales involved onto other players with longer time horizons and lower expectations for ROI. On-licensing returns revenues to the biotech company through a combination of initial payments (usually fairly small), success payments on subsequent development of the drug/medicine and a royalty on the final product. Thus, if the product does not get anywhere then the biotech company gets little cash. Usually when a biotech company closes such a licensing deal they announce it with a press release that quote the value of the deal as the sum of all the payments that they will get if the project is successful together with a plausible royalty flow: this however is the top of a range of possible revenue figures, the lower bound of which is nothing at all, and it will be spread over the next decade making its NPV much smaller. Companies that operate this sort of business through providing biology discovery, chemistry discovery, medicines development and medicines marketing have all been set up.

This business model is preferred by VCs, because it requires far less cash [189, 192–6]. The vast majority of 'biotech' companies funded by VC in the last decade have been companies whose goal is, or became, to fit into this drug discovery and medicines development path.

The most high-profile version of this were the genomics companies of 1995–2001 [194], which were going to radically overhaul the entire process making it faster and more efficient [197–8]. Enormous amounts were invested by public and private markets in US genomics companies in the late 1990s on this premise. Careful analysis of the genomics companies' business model shows that, even if you believe their own rhetoric on how effective their technology was, they still made around 8% NPV on investment [129]. The reason is that, no matter how clever the technology is on which the company is based, it is only one part of the process, and no single alteration in the process can overcome the inherent inefficiencies and costs of the process as a whole, as many other parts contribute to that cost and inefficiency [10, 129]. Rather, dramatic improvement in any one step makes a modest improvement in the whole process only – it requires the whole process to be re-engineered to make a radical improvement, which brings us back to Genentech and the FIPCO. Constrained by the need to fit into a conventional discovery and development process, the technology could not be unleashed to its full potential.

The same is true of many other discovery technologies that have received enthusiastic investor support over the last 20 years. Structure-based drug design, high-throughput screening, antibody technology, antisense and siRNA have all been 'the' solution to discovering new drugs, but in reality all are tools that fit into a single part of the process, but do not solve 'the' problem with the process, because there is no single 'problem' [199–200]. The stated belief of investors, and no doubt genuine but mistaken belief of entrepreneurs and scientists, that today's technology is *the* technology has been proven wrong repeatedly.

'Genomics' was the most high of high-tech, a solution that pharmaceutical companies were queuing up to buy, but it still could not make money. VCs might argue that 'the path to profit in the pharmaceutical industry has been well trod' [201], but it is not the path to the levels of profitability that VCs demand. It is merely the profitability of normal, real-world companies making products.

The obvious solution to this is not to try to use your technology to make new medicines, but to sell it to people who are making new medicines. This would then be a revenue-based business, selling actual product to actual customers. The large majority of the companies in industry directories such as the BioCommerce databases, BioScan, Nature Biotechnology Directory, etc. are in fact of that type – they are businesses making a product and selling it. However, they almost never get VC funding – they have to bootstrap their business from founder capital and profits. A good example is AbCam[2] whose own company history describes how they sought VC funding for three years before giving up: during the same time VCs invested in a large number of loss-making drug discovery and medicines development companies.

However, I have skipped a step in this argument in assuming that VCs wish to create companies that can become world-striding commercial operations. The reality, as we have discussed before, is that they do not. They wish to invest in a drug discovery company or a FIPCO, grow it to a substantial (but still loss-making) size, and then sell the shares to others with longer time horizons and less extreme expectations of ROI. This is the model on which virtually all VC investments in drug discovery or medicines development are made. The reason for this is that, as I stated in Chapter 2, 'the biotech industry' is producing investment opportunities as its 'product', not drugs or enzymes.

Ironically, it is the very characteristics of drug-discovery companies that make them such poor investments that make them good feedstock for an investment business. For a sales-based business the business value is a rationally calculated multiple of current sales or current profit. However, for a drug discovery company the capitalisation is a multiple of potential future profits. If these can be made to seem very large, the company value is very high. There is enormous room for engineering of the perceived value of the company which is not present if actual revenues anchor the valuation to the real world.

The few companies that do have revenues based on services or product sales and which receive VC funds are therefore pushed into becoming drug discovery companies: usually VCs only invest in revenue-generating biotech companies on the basis that the company will make this transition. The example of Morphochem is typical. They were set up in 1996 as a company providing chemical services for drug companies in return for revenue, based around some very clever chemical synthesis technology. They received VC funding, and encouraged (or driven – the level of coercion is never clear in such cases) by their investors in 1999–2000 changed their business model to one discovering their own drugs. However, they were a chemistry company, so could only do one part of the path described in Figure 8.4, so to be more credible as a nascent drug company they acquired a US-based biology discovery company, Small Molecule Therapeutics (SMT), which had also had a revenue-based business model. Morphochem (and its SMT subsidiary) turned to its own drug. However, by 2005 they had not managed to get any drug into clinical testing. In 2006 they were acquired by Biovertis: the combined new company had 45 staff, compared to Morphochem's staff of over 100 at its peak. The founder of SMT meanwhile had left, set up a replica of the SMT business model called Ricera (without VC investment), and by 2003 employed over 100 people and was profitable. Other chemistry service businesses set up at the same time as Morphochem such as Argenta and BioFocus managed to survive over the same period on revenue, despite quite hostile economic conditions. These are not isolated examples.

So selling products or services works, but is unpopular with VC investors. VC-invested companies are driven towards the drug discovery or medicines development model for reasons to do with the VC's business model, not their own. The reason that all 'biotech' CEOs in 2005 acted as if they are running speciality pharmaceutical companies, and downplayed the technology angle of what they were doing, is primarily because their investors told them to do so. [202].

The need to make the potential for future profits seem large leads to powerful drives to over-promise on the technology. To quote Al Kolb:

> Often the overhype [in a new field] comes from venture capitalists who want large, short-term returns. It has to be successful immediately or it's considered a failure. That doesn't make the science bad. Some things take a long time to develop, some never pay off. . . . Simple product development based solely on theory is risky and difficult. In the development of a new field of science it is going to be hard to predict what the impact will be or in what time frame.
>
> [203]

This forces companies to overhype their technology and its capabilities and then make poor decisions in development programmes to support that hype, or at least to not be seen to fail to support it until an exit can be achieved. As Ian Gowrie-Smith said of the investor-driven crisis at Skye Pharma in 2006, 'It is unfortunate that the short-term interests of a small group of shareholders can cause such a disruption to the development of a company where real achievements are often measured in decades' [204].

In conclusion, getting into developing your own drugs is a business model that is hard to defend on rational business grounds unless you are well funded to pursue radical technologies that the major pharmaceutical industry is unlikely to pursue. Indeed, this was realised early in the 'biotech era': John Walker, director of investments at Charterhouse Japhet, commented that

> [v]enture capital firms should seek products rather than research. If the products are more than a year away from the market, then we are not the right people. If sales are not £5M within three years, then, again, it is not for us', and said as a consequence virtually all forms of drug discovery and medicines development was inappropriate.
>
> (Quoted in [205])

Ever since investor analysts have asked what happened to biotechnology applications in areas such as industrial enzymes, concluding that these are good investments where technology can provide a quantifiable value-add, the market dynamics are clear, and if successful you can be selling product in a couple of years of start-up [206]. The answer, in Europe, is that the VC's own business model happened to them. Investment in any company applying chemistry or enzymology to non-drug applications has been almost impossible to obtain on the East of the Atlantic [112].

But driving companies to be drug discovery or medicines development companies is not the only problem. Investors also change their mind about once every year on what sort of drug discovery or medicines development company they want their investees to be.

If you were setting up a biotech, drug-orientated start-up in the late 1990s, then the thing your VC shareholders would have compelled you to do was in-license drugs from other companies to 'build a pipeline'. In 1997–8 in-licensing drugs was the vogue (as investors lost faith in science's ability to deliver, and forced companies to 'have a drug') [207]. The rise of excitement over 'the genome' and many political pronouncements on the subject means that in 1998–9 biology discovery (those 'genomics' companies) became all the rage, and any company coming along saying that they were doing chemistry or in-licensing was told to stop and acquire genomics capability [189, 192–4]. Thus, Chiroscience, a company that had built a strong franchise in a specific area of drug chemistry, bought Darwin Molecular, a genomics company 6000 miles away in Seattle, a substantial distance in space and in technology from their strengths.

Genomics' is not one technology – there are a slew of technologies that can be called 'genomics' technology. By 2000 having a technology was not enough, because the existing players has filled all the sensible 'one technology' niches [189, 208]. So successful players had to adopt 'bundling' approach of building a 'platform' of different, overlapping or synergistic technologies to deliver (in jargon fashionable at the time, now rather tired) 'functional genomics'. By 2001 all forms of discovery biology were out of favour and chemical discovery and early development were the thing [195], but companies were still told to focus on creating their own drugs, not on earning revenue (c.f. the Morphochem example above). By 2003 it was a 'hybrid' model – companies that could generate their own drugs and also have a technology platform that they could sell to others were the only investable proposition [209–10]. By 2004 the No Research, Development Only (NRDO) company was the code for a company that did none of its own research but in-licensed drugs for development, and was the only thing that could garner investment: Theil [196] lists 31 such companies in the US alone, receiving substantial investment in 2003–4. The industry had cycled back to where it was in 1998. For companies that had substantial research departments this means that research scientists were being 'downsized' substantially [211]. By 2006 the drug discovery company was back on the agenda [212], and the scientists were being hired again.

This is not in itself a problem, and changes in the economic climate are inevitable. The waves of VC fashion were quite unrelated to what the pharmaceutical industry actually wanted (remember that these biotech companies did not aspire to be FIPCOs – they were to sell their products to larger, stable pharmaceutical companies before they were near completion). Thus, in 2002–3 pharmaceutical companies were seeking development candidates to fill up their pipelines [196], so the one thing that investors can be sure of is that a company created to generate intellectual property that fit this profile will have no clients by 2006–7, when their products are ready to licence, as the pharmaceutical company pipelines will then be full of licensed products and what they will be wanting is discovery projects. And so it turned out to be, and valuations of companies in the early stages of drug discovery rose in 2006–7 to reflect this [212]. If you have enough cash to weather these changes until the fashion cycles round to support your product, or have a revenue stream to support you through the lean times, then you can make good when the market opens (for example, see [213]). The major US companies in 'genomics', which had garnered vast amounts of investment in the late 1990s, had the resources to acquire the products, patents and technology to allow them to move out of 'genomics' as a business in 2002–3, although they retained their technological abilities, and used their cash to acquire a conventional medicines development business [210, 214], or to morph into 'radical therapeutics' companies, as Human Genome Sciences did in its change to a regenerative medicine company [215].

But this cannot work in Europe, because European companies are not given the resources. Chronic underinvestment means that companies have no room to manoeuvre, no flexibility. So when the controlling shareholders demand that a drug discovery company becomes a medicines development company or visa versa, the company has no cash to do so and no resources to carry on what it was doing regardless. The combination of investor control, rapidly changing and commercially implausible business goals and underinvestment dooms the companies to almost inevitable failure, unless they do something drastic.

9
Investor Blockade of Business

As we have seen, VCs claim that they support their companies in a variety of ways, and even if the reality of time management means that this is an implausible claim, surely they would not go out of their way to sabotage an investee company? Of course they do not. But the effect of the way they invest is to achieve a similar goal. To explain this I will have to digress into a short discussion on share structures in VC-invested start-up companies.

Two mechanisms are used by investors to ensure that their shares gain a larger slice of any equity profit in a company than those of the existing shareholders. These are Preference Shares and Anti-Dilution Provisions (sometimes called Investor Protection Provisions). The *prima facie* rationale for these mechanisms from the investors' point of view is clear: they get more upside for the same investment, and hence a better return. However, the effects are the reverse.

I will discuss the effects of Preference Shares here, as my research has focused mainly on them. Liquidation Preference Shares ('Prefs' for short) provide their owner with a preference on the assets of the company in event of company liquidation. In the event of the liquidation, they get the Preference amount (usually at least the amount they invested) back before the other shareholders get anything, thus protecting their investment against the risk that they might lose money on their investment. This is a problem for any future investor in the company, as money they put in will (in the event of a liquidation) all go to another investor. So new investors in a company that has Preference shares will want Preference shares of their own, and usually will want ones with a prior right over assets (technically, 'superior' rights); so they get money first, the original investors get money second and anyone not holding Preference shares gets money last of all. The effect

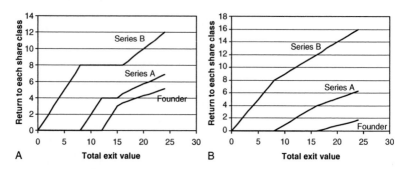

Figure 9.1   Preference share effects

Note: Example of the amount received by different share classes on liquidation of a company. Series A investors invest £4 million at a pre-money valuation of £3 million for 1x Preference shares. Then Series B investors invest £8 million at a pre-money valuation of £8 million for 1x Pref shares. Left panel – preferences are 'simple' Prefs, right panel, preferences and 'participating' Prefs. Shown are the amount each shareholder gets back (Y axis) for different valuations of the company at liquidation (X axis).

Source: Redrawn from [148] with kind permission of Journal of Commercial Law Studies.

of this on who gets what from winding up a company is illustrated in Figure 9.1.

Figure 9.1 shows an example of single, or 1x Prefs – the Prefs holder gets their money back before everyone else. This is in line with what used to be understood from the concept of a liquidation Preference share. But note the difference between Figure 9.1A and Figure 9.1B. In Figure 9.1A the shares are 'simple preferences'– the investor gets their money back, and then everyone is *pari passu*. In the much more common *Participating Preferences* (Figure 9.1B) the investor gets their money back, and then the result is split between investor and other shareholders. So under all circumstances the Pref shareholder does better. Pref shares are therefore used to 'protect' the investor even if the company is liquidated at a profit.

Academics give three reasons for VCs' requirement to invest in Prefs, those of tax efficiency [216], agency cost [217–18] and downside protection. The first of these does not apply at the moment in Europe, specifically the UK where Pref shares are less tax efficient.

Agency Cost is a term in the Principal-Agent theory, a framework for the analysis of the relations between investors and investees. In Principal-Agent theory, we consider the different motivations of someone who has assets (the Principal) and someone else who uses those assets to generate value (the Agent). In the biotech/VC context, the Principal is the VC and the Agent is the entrepreneur or biotech company management. The Agent's agenda usually is different from the Principal's, and they have more information about the operational realities of the planned business opportunity – in the terminology, there is *information asymmetry* between Principal and Agent. So the Agent may act 'opportunistically' to exploit situations with actions that are to their short-term benefit but to the detriment of the interests of the Principal, and to those of the company as a whole. Such behaviour is called 'amoral' in that it is not in the spirit of the original contract between Principal and Agent [219]. The problem for the Principal is therefore to set up a contract whereby the Agent's desire to maximise their reward for a given effort results in him or her also maximising the return for the Principal. If this is achieved, the Principal does not need to know what the Agent is doing day-to-day, nor to control them: the Agent's own self-interest will see to it that they do the best they can for the Principal [220].

The use of Prefs is seen in this context as a tool to get the Agent (entrepreneur) and Principal (VC) working towards the same goal. According to the agency cost explanation [217], liquidation preferences motivate the entrepreneur to achieve extremely high returns on capital necessary for the VC in this very risky investment field [71, 221–3], as they provide that the entrepreneur gets nothing if the company is moderately successful, and a rapidly increasing fraction of the company if it is very successful. The entrepreneur is therefore motivated to achieve the extraordinary results that the VC wants. In this theory, the entrepreneur's willingness to issue liquidation preferences shows that she thinks the company is worth more than the VC's liquidation amount.

In fact, Preference shares are not the only mechanism to achieve the same risk-reduction for the investor. Debt mechanisms can achieve the same goals – the entrepreneur gets nothing until the debt and the dividend/interest (which can be set arbitrarily high) are paid off [221]. In the financial literature, the two can be considered as equivalent (see e.g. Shirley [224]). There is an argument that debt is not preferred because of impossibility of determining the entrepreneur's commitment, and hence the chance that the entrepreneur will 'behave opportunistically' by taking project pay-offs available and default on debt (i.e. run off with the cash). But, in reality the veto powers I discuss in Chapter 7 mean that this outcome is impossible. Debt or Preference shares under extensive veto powers have the same effect [42], and there are a number of reasons for preferring debt as a mechanism of investment in private companies, in part to avoid the many conflicts of agenda and interest discussed in this book [224]. However, debt is almost never used by VCs: Gompers [62] reports that 96.4% of early stage investments are in preferred stock,

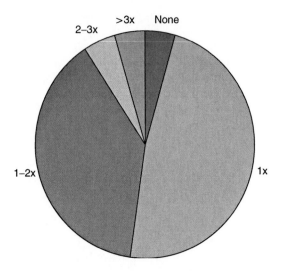

Figure 9.2   Preference multiples in 44 UK biotech investments, 1998–2005
Note: Number of investors investing in shares with different preference privileges in UK biotech companies, in this data set: none = no preference rights, 1x = preference over the amount invested, 1–2x = preference over between 1 and 2 times the amount invested (usually through guaranteed dividend mechanisms), 2–3x = preference over amounts greater than 2x and up to 3 times the amount invested, >3x = preference over more than 3 times the amount invested.
Source: Figure redrawn from [148] with kind permission of Journal of Commercial Law Studies.

2% in common stock and less than 2% in debt. The Principal-Agent theory falls down on this and other predictions it makes.

In reality, interviews with VCs demonstrate that Prefs are used to reduce risk [148]. But this is not just downside risk, which as we have noted can be defended against by loan mechanisms among others. Prefs are also there to protect the VCs profit. Two further aspects of Pref shares are used to 'protect' investors' 'upside' (i.e. profit). The first is the 'multiple Preference share' where the preference amount is a multiple of the amount invested. This means that the Prefs holder gets several times their money back before the ordinary shareholder sees any return (the multiple depending on the contractual terms). Figure 9.2 shows a typical distribution of terms.

Multiple preferences are the norm rather than the exception among biotech investors, and ensure that, unless the company does really badly (which we must confess happens quite often in biotechnology) then the senior Pref shareholder will get at least their money back. So the VC can get upside without downside, and pass the risk on to the

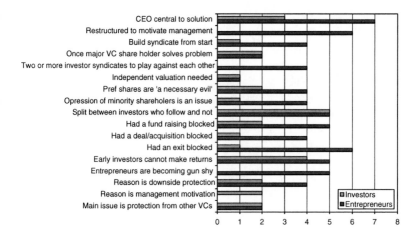

Figure 9.3   Causes and effects of Pref shares
Note: Summary of the issues and opinions expressed during interviews in this programme. Left-hand side – the class of issue raised as a pressing problem *for that interviewer*, i.e. a problem or solution of which they had personal experience in the companies they founded and/or managed. General classes of issues are grouped as 'Reasons' – what the interviewee's experience lead them to conclude was the reason for multiple layers of senior Preference shares, 'Problems' – what problems has been caused for the company's commercial progress that were directly attributable to the company's share structure, and 'solutions'– components of potential or actual solutions to the problem.
Source: Reproduced from [148] with kind permission of Journal of Commercial Law Studies.

management, founders or angel investors. This was confirmed by our interview study of investors and entrepreneurs as to the causes and effects of Prefs in biotech investments, summarised in Figure 9.3. (Our study comes to similar conclusions as those drawn by an excellent US study by Fried and Ganor [149]: the situation is aggravated in Europe by underinvestment, which means that the crises described below are more likely to happen in European companies than in relatively well-funded US companies.)

Few people on either side of the table saw entrepreneur–investor alignment as the main issue – the only reason offered for Prefs was upside protection and protection of VCs from the opportunistic activities of *other* VCs. Entrepreneurs were considered a minor nuisance in the imposition of Prefs.

To make this mechanism more effective, 'Prefs' have been expanded in VC usage since 1990 to include preferred treatment for Prefs holders in a variety of events, not just when the company is being wound up. Importantly, these include flotation (IPO) and sale of the company. This has substantial implications for the company, and allows the VC to gain shareholding in the company (at the expense of other share-holders) under a wide range of circumstances that debt instruments would not allow.

In Europe, the practical requirements of an IPO demand that the floating company has a single class of share, for reasons discussed in more detail in [148]. Preference shareholders will therefore have to give up valuable rights when the company IPOs, and this they are loath to do. The solution is to treat IPO as a liquidation event, and convert Prefs to ordinary shares as if the company is being wound up. The qualitative effect is that the ordinary shareholder or junior Prefs holder ends up with less of the company than their fraction of the number of shares that they owned, and can end up with none at all. (In the US this mechanism does not apply.)

Similarly, if the company merges or is acquired, the acquiring company will pay a particular amount in cash or shares to acquire all the shares of the purchased company, regardless of whether they are Pref shares or not. If this is divided among the shareholders according to the number of shares they hold, then the Preferred shareholders will have lost their rights over the ordinary shareholders. So on merger or acquisition the rights due to Preference Share are triggered as if the company were to be liquidated. A trade sale or other merger event was considered a 'liquidation event' in the articles of 14 out of 15 of the companies analysed in Maynard and Bains [148]. Again, this can result in one or more share classes being 'squeezed out' prior to a merger, as if the company was being wound up.

Anti-Dilution clauses can be even more malignant. These compensate the rights holder with new shares if the value of their shareholding falls. A 'down round' – an investment round in which the pre-investment value of the company is deemed to be less than the post-investment value at the last investment – can trigger Anti-Dilution provisions and result in the ordinary shareholders suffering all of the loss in value and the Anti-Dilution rights holders suffering none. However, the reason for that fall need not be an external event at all. It can be a notional or internal revaluing of the shares, most commonly a revaluation associated with new investment in a private transaction. Thus, holders of Anti-Dilution rights could declare that a company has fallen in value and issue themselves new shares under that mechanism if they controlled the company, to the substantial detriment of other shareholders. This does happen.

Both Preference shares and Anti-Dilution provisions are not unique to European biotechnology. Other industries and other geographies routinely use the mechanisms. But comparison with the US is informative. A 'down round' is always to be avoided, because it implies that a company's value has not increased. But the reason for avoiding it in Europe and the US is considered different *by the investors*. In the US a down round in a B financing (i.e. the second major VC financing) is to be avoided because it taints the company with a history of commercial failure, looses momentum in fund-raising and consequently demotivates all company's members. That a down round loses value for the investor is not the principle issue, as the investors believe that they can regain that value by subsequent work with the company [225]. By contrast in Europe the issues are more ones of VC PR (one European biotech VC was notorious for stating that they 'had never had a down round' in the run-up to raising a major fund). Pref share structures, Anti-Dilution clauses and the extensive control through veto powers and Board positions means that for the VC's value proposition a down round means little. As a result, while down rounds were common on both sides of the Atlantic in the 2002–4 downturn, in the US the effects on ordinary shareholders were substantially less than in Europe.

This is pretty rough on the European entrepreneur (who normally will only own 'ordinary' shares), but one can argue that they go into the transaction as a rational, informed individual and cannot complain when their dreams of wealth are not realised. However, the effects of Preference shares are more malignant, in that their existence can, and often does, block the company from potentially valuable transactions, including ones that could save it from the business impasses I outlined in the last chapter, including trade sales and IPOs. Given that VCs all drive their companies towards an IPO to achieve liquidity in the shares and access public market capital pools, putting in place a structure that prevents this happening seems peculiar to say the least. The case histories, however, suggested that it was not uncommon.

All 16 out of 16 interviewees in Maynard and Bains [148] with direct experience of trying to achieve exits for their companies stated that having a Pref share structure in place was a problem, and in general the more complicated the structure the bigger the problem. This is a problem not of the existence of Preference shares, but of the existence of multiple layers of Preference shares all with different thresholds and shareholdings. This situation is almost impossible to study analytically because each company ends up with a unique share structure, so we will present some case studies to show the problem.

Intercytex was founded in 2000 to develop therapeutics in wound care and cosmetic medicine. Over the next five years Intercytex received

£22.5 million from eight investors in three rounds, and was considered mature enough to float its shares in the public markets. The brokers suggested a value of the company in the range £20–5 million would be acceptable to the markets, which would have just about valued the investors' shareholding in cash, which was quite a good result for the time – several other companies of the same vintage had recently been acquired for less than the total investment in them. However, B and C shareholders had preference rights which meant that a £20 million IPO valuation would result in the entire company shareholding being signed over to them, mostly to C series shareholders. A and ordinary shareholders were not willing to acquiesce to this, and so after prolonged discussions the IPO had to be withdrawn. The public reason given was 'poor market conditions' [226].[1]

The solution in the case of Intercytex was to 'flatten' the share structure by converting all the Prefs to ordinary shares (each Prefs was converted to several ordinary shares – the original shareholders still got it in the shorts, but ended up with something as opposed to nothing). Intercytex emphasised that achieving this required a strong CEO backed by a strong, independent Chairman of the Board, but they had to get the agreement of the shareholders, and specifically agreement from the VC shareholders that relinquishing their Prefs rights was a better outcome than the likely failure to achieve an IPO implied by their previous failure.

Fried comments that in the US, companies have flattened the share structure, but generally this is prior to a new round of financing in situations, where the liquidation preferences in place far exceeded the value of the firm, so that the junior share classes, and in particular the 'ordinary' shares have lost all value as motivators for incentivising employees. So the new investor insists on a recapitalisation that flattens the structure, and then it invests in a new class of preferred stock which is now the only class of preferred share (Fried *per comm*). This seems to be rare in Europe, and it is openly admitted that, because of poor valuations coupled with Preference shares, management share schemes are useless at motivating management in Europe, perhaps reflecting the more antagonistic attitude that European VCs have towards founding management teams (see [22] for more comments on this).

Cyclacel illustrates a more malignant outcome. The company was acquired by Xcyte in December 2005 for a nominal amount, and raised $45 million (~£28 million) in the transaction. Hailed as a successful exit by the company, in fact the transaction meant that essentially all of the shareholding ended up with the C class shareholders because of the multiple Prefs they held, and there was no value left for early investors.

This was clearly not what the other shareholders wanted, but the C class shareholders had a dominant position on the Board and could make the deal happen, at least in part by simply using their veto powers to block any other deal. Fear of this type of 'solution' to the Prefs problem (having one dominant shareholder force an event that is rewarding for them but unrewarding for all other shareholders) is reported to be a major limitation on companies' willingness to accept investment.

The failure of a previous attempt to float the company in 2004 also suggested that it was 'this deal or nothing'. The flotation effort illustrates the problem of arguing your shareholder into the ground before you can agree to any deal – it took so long to agree to the price, agreements and trade-offs between stockholders that the short IPO window of 2004 had closed before the company could act [227]. The company stated that the float had been 'postponed' not cancelled [228], but as subsequent events showed there was no serious attempt to float the shares again. Microscience suffered a similar fate at the same time, also claiming that it had postponed its float [229] and also ending up being sold for an amount that was less than the VC cash invested, and hence in a deal that would have returned junior shareholders very little [230]. The issue here is not whether the company is worth a lot or a little, however (although one must wonder at the standard of due diligence of round after round of investors). The issue is that the paralysing effect of the share structure did not allow these companies to float at realistic prices when the market allowed, and so the business opportunity was lost.

Stout [231] describes several more case studies of acquisitions where junior shareholders have been left with minimal value because of the dominance of Preference shareholders. In the case of Arrow Therapeutics the dominant shareholder was not the most senior shareholder, and so the outcome was such that all investor shareholders got a return. The CEO is quoted as saying 'Therefore it was a satisfactory return for all the shareholders', illustrating just how irrelevant angel and entrepreneur shareholders (who were heavily diluted and made little from the transaction) are considered to be in determining Mergers and Acquisitions (M&A) options.

Companies that failed to achieve any exit are harder to document, because there is no reason for the specifics to be put into the public domain, and many reasons for those involved not to wish the failure to be known. The following are, however, typical, and are documented in more detail in [148]. One Cambridge-based company was approached by an acquirer in 2004. The offer was below the top Prefs threshold, so the B series shareholders would get their desired yield, the A series would make a loss, and the management and founders would get nothing. Not surprisingly, it was impossible to get agreement at the Board on this deal, and after six months of negotiation 'the [potential buyer] got bored and went away'. The company subsequently restructured, and was then acquired. In a second case the same thing happened, but the company could not get its shareholders to agree on four successive occasions when acquisition opportunities arose, and so the company remains independent (and loss-making). The company also raised three rounds of VC investment in five years over this time – in essence, therefore, for over six years the technical guru who founded the company spent most of the company's life arguing financing structures with current, future and potential investors. This is an inefficient use of management time, and is entirely due to VC investment practices.

The other major class of issue raised was the problem of 'opportunistic' actions by Pref shareholders: actions which benefited them but not other shareholder classes. As discussed above, the VC investors in a company effectively form a management group that has control over Board actions. They are in effect Agents, and can therefore manipulate the Company's behaviour towards their own interests.

How common is such behaviour? This is hard to know – business decisions are private, and even if they are recorded in Board papers (themselves confidential) this is usually only done in terms of decisions made, not options analysed. Several entrepreneurs interviewed said that they had been the 'victim' of such Prefs holder opportunism. An example will illustrate the issue.

Arakis is (in our terminology in this book) a medicines development company set up in 1999 and sold to Sosei, a Japanese company, in 2005 for £106.5 million, after raising £49 million of investment in three rounds from funds managed by nine investment groups and from business angels. The sale of Arakis to Sosei was seen as a major success for the UK industry, and indeed it was, coming at a time when exits of any sort were rare.

After several rounds of investment, institutional shareholders had layers of Preference shares which would mean that management would get nothing out of any plausible exit. To motivate the managers, a class of shares called 'incentive shares', with rights superior to all the others was created[2] in 2004, with a complex structure of four hurdles that gave an increasing percentage of the sale price to the management as the sale price itself increased, with a particularly sharp increase in management reward if an exit of £100 million was achieved. The effect of this on the value gained by specific classes of shareholders at different sales prices

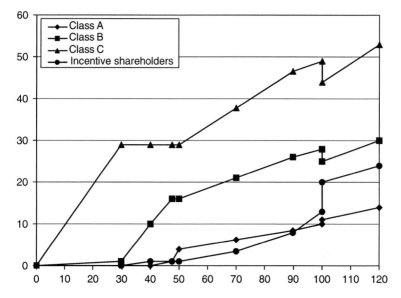

Figure 9.4 Arakis exit outcomes for major share classes
Note: Return to Arakis shareholders in event of a liquidation (from Companies House documentation). X axis – liquidation value of the company (=sale price). Y axis – amount returned to different shareholder classes. Vertical bars – hurdles for incentive shares granted to management ('Incentive/Ord').
Source: Reproduced from [148] with kind permission of Journal of Commercial Law Studies.

is shown in Figure 9.4. There is a clear jump in management return, and consequent drop in other shareholder return, at a threshold of £100 million. Arakis was sold for £106.5 million. This is an excellent outcome for all concerned, but lead the cynical in the industry to wonder, if the threshold of management Prefs options had been £120 million, whether the sale would have been for £126.5 million.

The examples above are primarily exits for the shareholders concerned – the original company has ceased to be an independent, private entity. Another, strongly promoted form of reorganisation is the merger, in which the original companies retain some part of their identity and shareholders are looking (it is argued) to build business capability rather than achieve liquidity for their shares. There has been strong argument for M&A activity in the biotechnology sector, especially in the last decade, which has seen a surge in formation of 'spin-out' companies which are widely seen as unsustainably small [232], and M&A activity has been rising over the last few years [195, 233–4], although more slowly in the UK than in the US [235].

In some ways, this is surprising, because there is a consensus in the financial literature that a strategy of M&A, and especially ones that involve a degree of diversification, result in loss of value and profitability, for a variety of reasons [236–42]: this specifically applies to the pharmaceutical industry [243]. The apparent success of many in the short term is due to accounting anomalies, short-term savings such as reducing R&D headcount and short-term stock response to newsflow [244–5], illustrated by short-term surges in the stock prices of Vernalis during their merger history. However, these studies are of large, public and usually profit-making corporations.

The problems of share structure discussed above should also apply to merging companies, but in this case there is a simple 'solution': if a single entity is a major shareholder in *both* merging companies, then they could drive a merger through by controlling the actions of both companies. Does such common shareholding remove the problems associated with Pref shares? In a sense, but they replace them with another problem, which is that mergers now are driven by the interests of the common shareholders, not of the company.

Maybeck and Bains studied this in a dataset of all European biotech mergers (i.e. joining of two companies of roughly equal size) between 1995 and 2002 [246–7] for which appropriate metrics were available, and found that indeed the large majority of merging private European biotechnology companies share at least one shareholder, as shown in Figure 9.5.

So it is clear that having shareholders in common helps to enable mergers to happen. But what are the motivations for the merger? A merger could benefit shareholders (by increasing their share), the company (by boosting its business, capability or both) or benefit both (by increasing the value of the shares). Setting aside the theory and rhetoric of what a merger is supposed to achieve, what is *observed* to happen when two European biotechnology companies merge?

There are two types of benefit we can imagine. As discussed above, biotech companies that are undercapitalised and driven by the vagaries of the business fashion of the year can find themselves in a technological or commercial cul-de-sac. A major corporate reorganisation could be the catalyst for change. Part of this change is the need for 'critical mass' – enough size to compete in an aggressively global industry.

Merging companies are on average more different from each other than companies picked at random from the 'universe' of biotechnology, although not much so [247]. This suggests that the merger to build 'critical mass' (i.e. taking two companies with similar or related

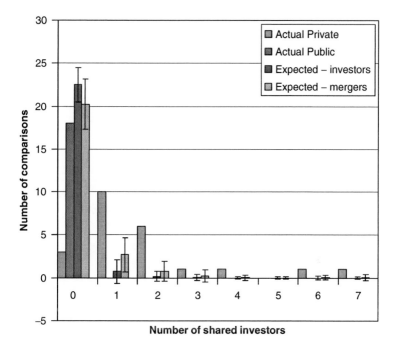

Figure 9.5 Shareholder commonality in merged companies

Note: Number of shareholders in common between merging companies. X axis – number of major (institutional) shareholders common between two companies. Y axis – number of comparisons in which that number was found. Categories: 'Actual' = data from mergers, private: private (Actual Private) and private: public (Actual Public). 'Expected' – what would be expected from random pairings of similar numbers of European biotech companies, either selected all European companies receiving institutional investment 2002–5 (Expected – investors), or from the companies that actually did merge. Error bars are 1.92 standard deviations above and below the mean for 400 selections of company sets for each of the two random sets.

Source: Redrawn from original publication [247] with kind permission of Nature Publishing Group.

businesses and building a larger one to address the same type of business) is uncommon. Rather, radical change – finding two companies that are completely different (that merge and continue to change) is more common, showing that change can be a major motivation. Thus, smaller companies tend to use merger to change something rather than to build critical mass. But, as discussed above, diversification has been found to be a poor strategic option for larger public corporations. Does change result in commercial improvement for biotechnology companies?

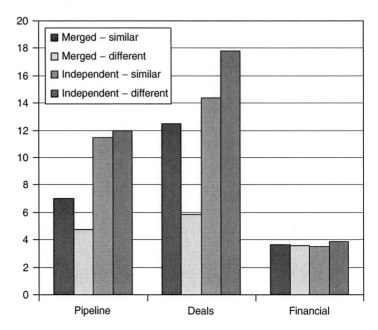

Figure 9.6   Commercial outcomes of mergers
Note: Fractional change in business parameters in merged companies and equivalent companies that have not merged. Categories: Pipeline (a score of their R&D product pipeline), Deals
(a score of their ability to attract major co-development contracts with pharmaceutical companies) and Financial (log share price increase). Shown are the average increases in these score
for small companies that merged and a 'control set of otherwise similar companies in terms
of age, size and business area which remained independent, separated into companies that
had similar business models at the time of merger and ones that had different ones.
Source: Redrawn from original publication [247] with kind permission of Nature Publishing
Group.

Figure 9.6 analyses this for three things that are important to the
commercial success of a biotech company: their product pipeline (i.e.
how successful they have been in discovering and developing products),
their deal flow (i.e. how successful they have been in closing contracts
with major pharmaceutical companies to take their products on to
commercialisation) and their share price.

In summary, merger does not affect the share price significantly, and
if anything has a deleterious effect on the other, more direct measures
of company performance. Biotech mergers are *observed* not to benefit
the merging businesses.

So if companies do not merge to gain critical mass, and do not
improve their business performance by merger, why do they do it? The

specific reasons for any merger are of course hidden in the confidential discussions of the company's shareholder and Board meetings, but the only logical conclusion from the analysis above is that, if merger does not benefit the company as a whole and the company is controlled by a subgroup of specific shareholders, then merger must benefit those shareholders. The aim of a merger was not to put together two *businesses* but two *companies*, to pool investor bases or (less charitably) to provide a common investor base with a mechanism for improving their shareholding in an investment climate that does not allow increase in their share price. Again, this is about biotechnology companies as the feedstock for a business in share trading.

The mechanism for doing this is again the Preference Share or the Anti-Dilution Clause. If merger is considered a 'liquidation event' then an Investor Shareholder can block any merger that is not favourable to them, regardless of the number of shares they hold, block any other business option through veto powers and so force a merger that triggers their Preference Shares at the expense of the shareholding of other shareholders. The drive for a merger is therefore not that it increases company value, but that it increases shareholder value for the Pref shareholder by reducing the shareholding of the other shareholders.

The merger of Amedis Pharmaceuticals and Paradigm Therapeutics is illustrative, and entirely typical. This was the merger of two private companies to make a third, private company (Amedis ended up as a wholly owned subsidiary of Paradigm), so this was neither an exit event nor winding up of either business. However, this was an event which triggered Preference rights (and in this case 'Anti-Dilution rights'). If the value of the transaction was deemed to be sufficiently low, then the value of junior classes of share would be reduced to zero, and so the acquired company's ordinary shareholders would not receive any shares in the acquiring company. As the founder and 'business angel' shareholders in Amedis disappeared from the share register of the merged company, while VC investors did not (and neither did the personal shareholding of the VC managers), we can presume that the value put on Amedis when it was acquired by Paradigm was lower than the sum of the Preference amounts.

(One might ask why the founders and business angels allowed this situation to occur in the first place. Why did they accept such clauses? We discuss this in Chapter 13.)

In this argument that investors dominate the nature of M&A activity in private biotech companies, it is worthwhile to dispose of another myth of the industry at this point, mentioned in many articles on business transactions of the type discussed above – major collaborative programmes, acquisitions, mergers and IPOs – which usually cite the unrealistic expectations of the management or founders, or their 'ego' (i.e. vested career interests other than cash) as the principle limiting factor in closing the deal. Thus, Esposito cites management egos and management denial of the scope of the problem their company is in as the principle blocking issues for public companies which would benefit from a radical change of direction or structure [248]. Colyer [249] quotes the following exchange in a management forum:

> At the BIO CEO & Investor Conference this past February (1999), Ben McGraw, chairman and CEO of Valentis (Burlingame, CA), started his talk on biotech industry consolidation with an interesting exchange. To a room filled mostly with senior management from the biotech industry, he asked:

> > *McGraw*: Who do CEOs work for?
> > *Group answer*: The shareholders.
> > *McGraw*: Is M&A in biotech good?
> > *Group*: Yes.
> > *McGraw*: What is the primary deterrent to M&A?
> > *Group* (in unison): CEO ego.

Amihuv and Led find the opposite, that mergers encouraged, even driven, by the need for management to secure their jobs [250]: however, this is in major conglomerates rather than individualistic start-ups. The more general conclusion from all such studies is that whether a merger happens or does not happen is dependent on the desires of management, not shareholders.

We have examined the fallacy behind the assumption that M&A is good for the industry – appropriate M&A might be, but *as it is practiced* it is in fact not good for the industry. However, my analysis above also suggests that the 'ego' argument is also spurious, at least for those early companies that are effectively run by their VCs. An editorial in the same journal recounts stories of two companies whose strategic moves were blocked by VCs. In one case a sensible chemistry company acquisition by a cash-rich larger company was blocked because it would have crystallised a down round by the VC in the chemistry company. In another a German company was forced to close down rather than be merged because state-funding rules made that a better deal for the investors [232]. Stout also found in several detailed case studies that management was never obstructive, and usually highly constructive, in enabling an acquisition to occur [231].

The reality is that, of course, management careers are on the line for mergers, and this will motivate some of them to push for unwise mergers or oppose sensible ones. But shareholder issues dominate merger and other critical business decisions. To the extent that managers can oppose or drive through major strategic change in a biotech company in the face of share structure and extensive shareholder veto power, they do so because they are also (minor) shareholders or option holders in the company. But managers and entrepreneurs have little genuine power. Stout [231] also finds management teams relatively powerless to influence merger options and partners. And once a manager is 'out' they cannot defend their position, and their equity in the company is effectively lost. This is one of the reasons for the forced changes of management which I discussed in Chapter 7: removing management from a position of responsibility also removes any residual control they have over the company's fate.

10
Pomp and Circumstance

Having given the company too little money, fired the management, blocked business deals and driven the business into a commercial cul-de-sac, one might have thought that the VC/manager would at least want to conserve every penny of cash that the company had. However, the business model as it is pursued by VC-backed biotechnology companies in Europe does not appear to do this either. What I called 'Cargo Cult Economics' in Chapter 1 appears to apply here too – the argument that if big, successful companies do something, then if we do it too we will be a big and successful company. I will exemplify this with three examples of needless expenditure that VC-backed firms pursue in the pursuit of the illusion of being a 'big, successful company'. I should emphasise that generally these policies are pursued by companies that are *not* big, successful companies, and in some cases are not big or successful because they have run out of cash.

The scale of this cannot be overemphasised. This is not just a VC insisting on a new logo and a new set of office chairs in order to 'look professional'. Åstebro [251] surveyed a large number of Canadian inventors whose inventions did or did not receive VC investment. Only 6.5% of inventions reached market, but gross margins of those that do were 29%, similar to pharmaceutical industry. The paper analysed the returns to the inventing group and to investors (independent 'angels' or VCs investing in established companies). He found that the independent inventions survived on the market for a similar length of time, and the gains returning to the inventor (individual or corporate) were comparable. However, the development costs for successful inventions made by individuals were one-eighth of those for established companies! To an extent this is because the lone inventor focuses on simple inventions – new garden tools rather than gene therapy. But it is also because the lone investor focuses on adding value to their invention rather than building a corporate infrastructure.

To illustrate the problem with some specifics, I will look at three areas where VC investment drives companies towards unnecessary and unproductive expense: the Science Park, Management again and 'Process'.

'Science Parks' are property investment ventures designed to attract tenants with new, high-tech businesses. They are usually close to a major research university, although in reality there is often little interaction between the research centre and its Science Park. Cambridge claims to be the most flourishing centre for biotech in Europe, and has five or six Science Parks (depending on definition): the Cambridge Science Park on Milton Road, the Cambridge Business Park on Cowley Road, Granta Park, the Babraham Research Campus, the Cambridge Research Park in Waterbeach and arguably the Great Chesterford Science Park and Melbourn Science Park. (The whole 'cluster' is about 15 miles long along its long axis, about half the length of the Research Triangle Park, North Carolina.) But its only formal relationship with Cambridge University before about 1996 was that Trinity College, a component of the University, owned the land on which the original Science Park was built.

The Science Park phenomenon is an aspect of the 'cluster' effect. High-tech companies arise in localised 'clusters'. What a 'cluster' is, depends on how you chose to define it [252]. Cambridge, London and Oxford are spoken of as separate 'clusters' in the UK, but are no further apart than San Francisco and San Jose, which are part of the one Bay Area 'cluster' in the US. Medicon Valley bridges between Denmark and Sweden in Scandinavia, is considered one cluster. The intense parochialism of some European countries' politics is a strong negative force at work here, and we will discuss this in Chapter 12. But the effect is to make some addresses more desirable than others, and (in part because of this) much more expensive. Thus, VCs will prefer to invest in a company with a Cambridge Science Park address rather than one in Luton, a mere 15 miles away (and more conveniently situated for travel to London), because of the prestige associated with the address.[1] Companies therefore get set up on or move to Science Park addresses, and have to pay rent accordingly.

Lindelhöf [253] has analysed this effect in Sweden, to see whether in reality the science park added value to companies. A major difference was that on-park companies were much more likely to be VC funded, although the reliance on external capital was the same for on-park and off-park companies. The only other general theme was that on-park companies had higher employee education levels at all grades, consistent with the 'park as people magnet' hypothesis. But overall investment (from angels and public markets as well as VC) and company success was the same. The principle effect, then, of driving a company onto a Science Park in a prestigious cluster is to put its property costs up, and to enable them to raise capital to cover those costs.

Surely a prestigious Science Park address helps with relationships with customers as well? I have not found any data on this, but anecdotally I can say that in three years at Amedis, during which time we negotiated with dozens of companies and closed three major collaborative R&D programmes, at no time did the client pharmaceutical company visit our offices. They were interested in the product, not the postcode. Informal discussions with several other Cambridge-area start-ups have confirmed this. Customers do not care where the company is, they only care what the product is. Amazon.com proved this in 1995.

I have discussed the tendency to replace the founding management team by a new, more 'professional' team, and how VCs explain this by explaining the primacy of management as an investment criterion. Leaving aside that the existing new venture team, the people with the vision and drive to create the company, might still have something to contribute, who are the people who replace them? The usual answer is, *from the standpoint of a start-up company*, that they are mediocre and very expensive.

The typical new management hire has experience running a major R&D group or division of the pharmaceutical company. They may be star performers in their old jobs, but in the context of the start-up they are not what is needed, in terms of performance, skills and experience or salary expectations. A fairly crude analysis illustrates that having a Big Pharma employee as part of the top management team does little for market capitalisation of new biotechnology companies and suppresses company growth or investment (Figure 10.1). Having top 20 pharmaceutical experienced executives in your team makes marginal difference in capitalisation, funds raised or the ability to attract funding rounds or investors. It does, however, slow down company growth substantially. Interestingly, having top ten biotech company experienced executives in your team seems to have no effect.

They are, however, what the VC expects a senior manager of a big, successful company to look like, and so in another bout of Cargo Cult Economics they appoint them hoping that the companies will become big and successful as a result.

The company these executives come from depends, as one might expect, on the country where the recruiting start-up concern is located.

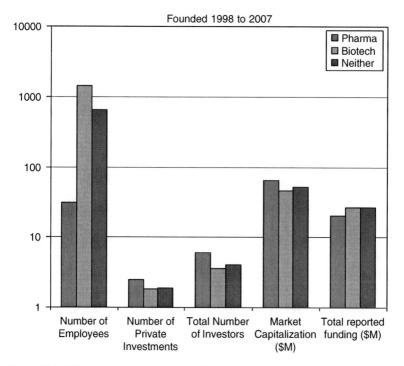

Figure 10.1 Big company experience and biotech success
Note: Some measures of company success, analysed by whether a company has at least one person with experience of major pharmaceutical companies (Pharma), major biotech companies ('Biotech') or neither in its executive team (Exec).
Source: Data analysed by the author from CapitalIQ (www.capitalIQ.com) for 193 companies in the UK and 116 companies in Germany.

German companies tend to recruit from German companies, UK companies from companies with strong operations in the UK (Figure 10.2).

This suggests that the biotech executives, many of them CEOs, recruited into biotech companies under VC control are not the globe-striding commercial giants they are made out to be, but follow the same recruitment rules that we saw in Figure 4.5: indeed, if anything previous 'big company' employees are more conservative about moving to new opportunities than the average.

Surely an executive with 20 years' experience of taking drug candidates through from idea to product, managing budgets, building teams and convincing the powerful and the well-clothed of their success, will be better at it than someone who has not done it before? This may be true. But this is not what they are being hired to do in a start-up company. The job

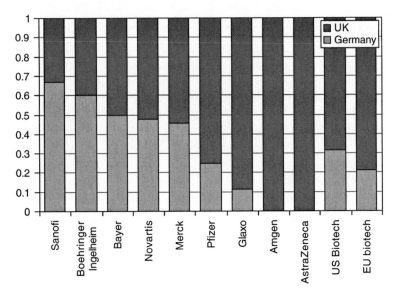

Figure 10.2   Where do Pharma-experienced biotech execs come from
Note: X axis – which pharmaceutical biotech company with major European operations did European biotech executives with pharmaceutical company backgrounds come from. Y axis – fraction of the executives with these background that go to UK or German companies. This analysis uses the same dataset as Figure 10.1.
Source: Data from CapitalIQ.

there involves a seat-of-the-pants combination of creativity and business development and, as we have seen, intense negotiations with potential funders, probably for at least half of their time. In many cases, the recruits' previous, big organisation job does not involve business or technical expertise, but rather the ability to function in a big company management structure. This is a very different task from creating a new company which, if it follows the visions of its founders, will probably be building a business that no one has created before, and if it follows the visions of its investors will change its vision at least once a year. This, all on a budget that is two-third of what they need to achieve it (about one-fourth of what they are used to), is not what they are trained for.

This is made worse by 'Anchoring' – planning for the future based on the optimistic assumptions built into the original plan, in this case the 'business plan' that the entrepreneur wrote to convince the VCs to invest. The entrepreneur might understand the limitations of the plan's assumptions. The incoming CEO does not. Anecdotally I have seen start-up CEOs take a single speculative line in a business plan as if it were the established truth and spend some scientist-years of research on a fruitless line of research as a result, despite being told by the founding Chief Scientific Officer (CSO) or Scientific Advisory Board (SAB) Chair that failure is almost inevitable.

I must emphasise that this is the *statistical* outcome seen from paying ex-pharmaceutical and ex-biotechnology company executives to join the top ranks of a start-up biotechnology company. It shows that the argument that Dr X is a better CEO candidate than Dr Y *because* Dr X has a Big Pharma background is fallacious. This does not mean that Dr X *is* worse than Dr Y. Dr X may be a superb candidate, with a clear grasp of operating in a fluid, cash-strapped environment, of the need for him personally to buy stationery and organise cut-price air travel, a brilliant presentational style and so on. I have known several ex-large-company start-up CEOs who were excellent at their job. But it is a fallacy to say that the NVTMs cannot be executives in the company created because they do not have the name of a major pharmaceutical company on their CV.

Having said this, it must also be said that entrepreneurs are not the most balanced and rational of people either. They are nearly always more optimistic and overconfident than non-entrepreneurial managers [254]. There are complex reasons for this, but most of their overconfidence is *not* the result of ignorance or inexperience, but of the incurable optimism that lead them to create a company in the first place. However, Figure 10.1 shows that, no matter how arrogant, optimistic or in some cases downright weird they are, they get the job done as well as teams dominated by ex-big-company middle management.

We might also note in passing some VCs' predilection for MBAs in the recruitment process, possibly because the business education is seen as a counterbalance to the world-class science education noted in Figure 4.5, and for CEOs with financial backgrounds that resonate with the VC's own interests. Is this a good thing? Top-tier MBAs command even more substantial salaries than other big company middle management, but as Paul Graham noted [255],

[i]f you work your way down the Forbes 400 making an x next to the name of each person with an MBA, you'll learn something important about business school. You don't even hit an MBA till number 22, Phil Knight, the CEO of Nike. There are only four MBAs in the top 50. What you notice in the Forbes 400 are a lot of people with technical backgrounds. Bill Gates, Steve Jobs, Larry Ellison, Michael Dell, Jeff Bezos, Gordon Moore. The rulers of the technology business tend to come from technology, not business. So if you want to invest two

years in something that will help you succeed in business, the evidence suggests you'd do better to learn how to hack than get an MBA.

Of course, business qualities are critical. What this quote illustrates is that the qualities selected for in the MBA admissions process and taught in the course are not the ones that are associated with success in high-growth new businesses. Similarly, in a more systematic survey Andrews showed that CEOs with a financial background do not achieve better stock performance than non-financial ones [256].

There is one major exception to this, which is that VCs are willing to allow so-called serial entrepreneurs to remain in charge of their start-up companies in Europe. There is a touching belief that entrepreneurs who have started companies that are subsequently successful are almost inevitably going to be successful again, a belief that the entrepreneurs themselves foster, of course. There is one sense in which this is true in biotechnology, but in general serial entrepreneurs do not show any better performance in company success than novices, providing the novices themselves have business backgrounds and goals. To quote Birley's extensive research in this:

> There is no evidence from (the [257]) study to suggest that those new businesses established by 'habitual' founders with prior experience of business venturing are particularly advantaged compared to their more inexperienced counterparts. Contrary to myth, it would appear that the majority of 'business engines' do not appear to create second-time-around ventures which out-perform their 'fledgling' counterparts in terms of job generation and wealth creation.

In other words, 'successful' teams are lucky, not talented. Westhead also found no difference in performance between firms founded by novice and habitual entrepreneurs [258].

The one sense where successful serial entrepreneurs are at an advantage over novices is that their companies are better funded. Westhead found that the one difference between novice and serial entrepreneurs is that the latter brought more personal wealth to their companies [258]. And if VCs believe they will succeed, they will invest more in them, and as we have examined before this will allow them to succeed. Investing their own money in the enterprise is also seen as inherently worthy – they have 'hurt money' or 'skin in the game', showing their commitment in the only terms that VCs can objectively evaluate. Again, the amount of 'hurt money' invested by the entrepreneur founder is not related in any way to the financial outcome of the company years later. Busenitz [259] looked at exactly this point in 183 VC-backed ventures and found no correlation between entrepreneur investment at the start and outcomes. (We should note, however, that it might alter the outcome for the *entrepreneur*, by reducing dependence on VC investment.)

The third aspect of Cargo Cult Economics that I will touch on in this chapter before going on to ask why VCs do all these extraordinary and counterproductive things, is what I have called 'countable features'. VC managed businesses tend to be pushed towards accumulating features that can be counted. Two examples will suffice to illustrate this.

Patents are widely agreed to be the backbone of business in biotechnology, as we have discussed. While this is not actually true of many companies working in the field, which rely on proprietary reagents, trade knowledge, expertise or just the willingness to go and do the job, it is true of pharmaceutical businesses that a secure IP position is critical to licensing a new drug to a development partner. Patent due diligence is a critical part of the investment due diligence process.

Thus, companies seeking VC funding are advised to file patents. Generally, the more patents filed the better, but as the amount of due diligence done on those patents is small, at least for private funding rounds, these can be purely 'bulk' in the patent portfolio. Three patent agents working for major UK agencies have said to me that they are frequently told by client biotech companies that they need some cheap patent filings just so they can list them in their business plan in the section labelled 'Intellectual Property'.[2] However, even these business-plan fillers cost money to draft, file and maintain. Critically, they cost a relatively small amount to file but an increasingly large amount to prosecute and maintain, so filing a patent is a commitment to expenditure of £50,000 or more in the future. They can also jeopardise future intellectual property options by prematurely or incompetently revealing inventions. One company of my knowledge was turned from a profitable business to a loss-making one by VC insistence that they 'file patents', even though their unique competitive edge arose from the expertise and skill of the scientific team, not from patentable technology. Having employed scientists to do research to create patentable technology, they were then dependent on investment for survival, and had to sell out to a larger company to avoid running out of cash. This is, regrettably, all too common. A second company spent an estimated £50,000 on filing and pursuing 24 patents, only to run out of cash or enthusiasm for all but four, in the process making sure that the ideas embodied in those patents could not be patented by anyone else (because the patents had

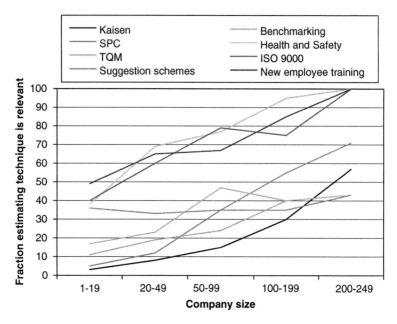

Figure 10.3 Management techniques used by SME technology companies
Note: Data set covers all types of companies.
Source: Data from [159].

been published). This is at best a waste of time and money, at worst a waste of good ideas as well.[3]

Management techniques are a second countable value that some VCs are keen on, from formal employee manuals and performance review processes (with associated HR manager) to ISO 9001 certification (and associated Quality Manager). I have seen several start-ups have to compile lengthy Employee Manuals as soon as they received VC funds, despite only having ten employees, three of which are directors. This costs money – usually in such cases several man-months of specialist HR consultant time. VC-backed companies are more likely to have VPs of Human Resources (HR) than non-VC-backed companies: VCs prefer HR role to be conducted by someone identified by a title. The presence of a VP of HR (or equivalent title) in a company has no effect on stock performance post IPO [256].

Performing research and development to the highest standard is absolutely critical, and we have seen how failing to do so can kill a company [69]. But investing in this is not the same as investing in those processes suitable to maintain quality in the routine jobs found in large manufacturing organisations. In general, formal 'Management techniques' – the approaches typically learned by large company management, sold by consultants and requested by VCs, do not add to the value of small companies [159]. As firms grow their use of formal management techniques increases, but the only ones used even by more than 60% of large companies are ones which are backed by regulation or law, such as ISO 9000, Health and Safety procedures and basic employee training (Figure 10.3).

North [260] found that implementing BS 5750 (which was the UK precursor of ISO 9000) in small companies was actually harmful: small companies forced into doing so by clients or investors avoided the harm it would do by simply ignoring it after they had 'implemented' it. Those that used it did so as a marketing tool rather than as anything relevant to the operational aspects of their business. Hongi and Cheng also find that smaller companies only implement such management processes at the demand of their customers, and that it contributes little to their business compared to its impact on large companies [261]. Deakins [80] similarly comments that at best formal training programmes for small company management have minimal effect on company performance. Storey [262] analyses the literature and finds that, while the managers who go on general management and financial courses (and presumably their investors who pay for them in cases where that is relevant) say the courses are useful, econometric measure of their impact shows that formal training does not improve small company performance. We noted before that a lot of the 'professional' advice that small companies receive is overpriced and not very useful (Figure 7.5). Specific, technical training, awareness and contacts are useful, but the sort of training that leads to management CV bulking is not. So imposing these types of procedures on small, private, research-based (and hence customer-free) start-ups at best does them no good, and at worst damages them: the only reason for doing it is PR for them or (more plausibly) their investors.

11
The Real VC Business Model

The previous chapters have outlined how VCs invest too little (for reasons that do not stack up), and then drive companies to a series of adverse decisions and business outcomes while failing to support the management they have emasculated with contractual arrangements. They appoint the wrong people, chose the wrong business strategies and block the companies' attempts to do anything about it. It is of course a one-sided story: there are many factors contributing to Europe's lacklustre biotechnology story, and many factors contributing to the few stars in the continent's dim sky. But the sections above show that VCs both control the companies and act so as to reduce their chances of success, through underinvestment and poor management. Their influence is so all pervading and their role in the industry so crucial that they must be a significant factor in its failure, and we have identified some of the mechanisms through which this occurs.

But this begs a critical question. VCs are not motivated by malice or political aims or (as is the case of many company founders, alas) the desire to see some beautiful science pursued at any cost. They are in it for the money. It is in their interest not to do things that will lose them money, and not to do things (if they have this option) that prevent them from gaining them money. Surely the logical flaw in blaming VC for the failings of the industry, no matter how the data appears to link them to industry failure, is that they have no motivation to fail.

This chapter examines this argument, and finds it wanting. Due to the baroque nature of what is called VC in Europe, VCs have no particular motivation to make their companies succeed, and in some cases positive motivation to make them fail.

The concept of VC is a US invention, and dates to the 1960s. By the time biotechnology came along in the 1970s VC was a well-established business in the US and Canada, although, as we have seen, it followed rather than lead the charge into the new industry. However, the explosion in VC in the US dates from the early 1980s, coincident with the biotechnology era, and just to be absolutely clear about this, the US biotechnology industry would not have grown at anything like the rate it has without VC. As we have discussed, European VC appeared at about the same time, although risk investment in technology companies had existed in Europe for some time before, just not trading under a 'VC' marque.

There have been many studies of how VC works, almost all analysing the US industry. It is useful to summarise the consensus view here, and then see what actually happens in the European biotechnology arena. See [71] for a more detailed description of the general US VC industry.

Typically, a VC fund is a limited partnership, a separate legal entity with two types of partner: General Partners (the management group acting to make investments, monitor them and sell them at the appropriate time) and Limited Partners (outside investors providing the money for investment). In the US there are tax advantages to this structure, and the same structure has generally been adapted to Europe. In the UK funds are often formally based in offshore corporations, again for tax reasons. In Germany institutional investors such as pension funds are not allowed to invest in such offshore vehicles (to prevent investment vehicles avoiding tax and regulatory requirements which the German government wants its pension funds to adhere to), so many non-German VC funds have onshore companies 'mirroring' them in which German institutional investors can invest.

The money for investment comes from outside investors. Usually the General Partner will put in a small amount of their own cash (e.g. 1% of the fund): the rest is from the LPs. Most funds (three-quarters according to Sahlman [71]) have a fixed life – all the assets must be realised and the money returned to the partners within the life of the fund. Partners usually do not put all their cash into the fund at once – funds are 'called down' as and when the fund needs them to invest, so the LPs have the use of the money until it is needed for investment. However, payment back to the Limited Partners usually only occurs at the end of the fund's life.

As the capital for investment comes almost entirely from the Limited Partners, the General Partner must have some operating capital to run the investment process, and some motivation to invest effectively. The former comes from a management fee, paid continuously throughout the life of the fund starting as soon as the fund is created ('raised'), the latter from a profit share, usually called a 'Carried Interest'. The Carried

Interest is there to motivate the General Partner to invest well. This type of reward structure is believed to 'work' in the sense of motivating all parties to the goal of maximising equity return. Studies of hedge funds, which are relatively unregulated and usually have the structure (at least in the US) of a managing partner providing management services for the fund and Limited Partners providing investment funds, show a positive correlation between the performance-based component of the management reward scheme and fund performance, following the VC model [263] (although, given that the hedge fund business in 2004 was $1000 billion, it is more accurate to say that VC mimics hedge fund economics rather than vice versa). Incentive-based funds take more risks than non-incentive-based funds – as the object in investing in VC must be to obtain the upside that comes from investing in early stage, high-risk companies, this should also point to a bias towards performance-based reward for the management team.

The theoretical business model for VC investing in biotechnology growth opportunities is therefore this: raise a fund, use a small fraction of the fund to support on-going operations, invest the rest in young, underappreciated companies, manage those well, exit them and gain reward in a profit share. Sahlman [71] estimates that a VC should expect to make at least four times as much from profit share as from management fees on an annualised basis. If successful at this, the VC (or its younger partners and managers) can then raise another fund based on their reputation, and repeat the process. The older partners retire rich.

In the US 'VC' has meant a fairly consistent type of investment since the 1960s. In Europe it has not. A wide range of funds have been created since the late 1970s on a 'VC' model, with the aim of investing in a relatively small number of early stage, high-risk companies which are then aggressively managed to yield very high returns. However, in Europe there has been a systematic trend for VC to move away from this to a 'merchant capital' model, investing more passively in much more developed opportunities [167]. This is a general trend for VC everywhere – it is not just a European phenomenon, and only North America retains VC *as* VC, rather than as a rather arcane form of private banking. Thus, by the late 1990s the large majority of 'VC' in Europe was concerned with management buyouts and buy-ins and mezzanine financing.[1] In the US this is called 'commercial banking' – the term 'venture capital' is reserved for much earlier stages of activity.

There are two reasons for this. Firstly, as noted in Chapter 2, early stage VC in European biotech makes very little money. Across all industries in the UK returns on early stage investment are poor: 2.6% annual return on investment for early stage investment and 3.8% for 'development capital' versus 19.4% for MBO/MBI business [167]. But secondly, and more importantly for the General Partners, they also generate far smaller funds and hence far smaller management fees. It is this issue of management fees which drives the business model of European VC.

As I noted, the theoretical argument as to why VC is motivated to build great companies is that the majority of their reward comes from their share in the profit made from equity in those companies. Sahlman [71] states that a General Partner in the US gets a management fee that is typically 2.5% of committed capital and a profit share (Carried Interest) from 13% to 30%. He calculates that this means that the typical salary + bonus for partner is $250k–$350k per annum, whereas the Carried Interest annualises to $200k–$1 million per annum. These figures are considered typical: Gompers [12] estimates the normal range of management fee is 1.5%–3% of funds under management and a 20% profit share (nearly all of their sample has a 20% profit share, around 5% of the contracts they examined stated a 25% profit share). This is rather more generous on the management fee than hedge funds (which we mentioned above as being an interesting comparator): these tend to operate a 1–2% management fee, and a 20% Carried Interest [263], but basically the structure is the same. In Europe the Carried Interest in funds is usually disbursed to all the fund managers on a pre-agreed ratio (i.e. the large majority to the Partners in the General Partner company) at the end of the fund. By contrast in the US Carried Interest received by an individual in a VC firm is more often linked to the investments that the individual has been responsible for identifying, making and managing. Thus, unlike in the US, in Europe there is no competition between 'my companies' and 'your companies' within a fund, but equally there is less incentive to drive 'my companies' to succeed providing that I believe the fund as a whole will generate money.

The argument VCs make when out raising funds is therefore that they will use the management fees to aggressively drive companies to high valuations, because it is in their interest to do so.

But in reality, how much do the European VC management groups make? Figure 11.1 examines how much the VC General Partner can be expected to make from a typical VC fund structure as a function of how much profit the companies themselves make.

(In this model there is a threshold profit below which there is no Carried Interest – this is a fairly common theme in European VC, analogous to the Preference Share structure that they in turn impose on their investee companies).

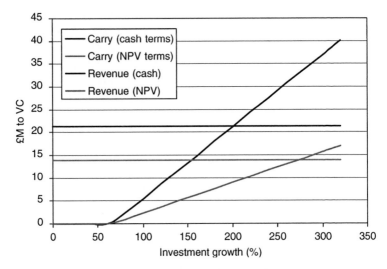

Figure 11.1   VC yield from management and Carried Interest
Note: Model of the amount of revenue to General Partner as a function of the return on investment from the fund(s) they manage. Shown are cash and NPV values of returns generated by management fees and Carried Interest.
Source: Author's own model.

This shows that the profit share is a minor component of reward unless the fund shows at least 150% *growth* (i.e. worth 250% invested cash), and Carried Interest does not become equally important as management fees on an NPV basis until about 300% growth is achieved. This means 300% growth *overall* on the fund over a 10-year timescale, that is, an annualised rate of around 12%.

Twelve percent is a plausible figure for VC overall: many generalist, late-stage European VC funds claim returns of at least this much. However, as we noted in Figure 3.2, it is almost never achieved by European biotechnology funds. This is not unique to European funds: profit payouts to Australian funds only matched management fee payouts in the extraordinary stock market bubble of 1999–2000 (and even then they did not exceed them): the following year funds received only 3.8% of their total reward from profit share [264].

In addition, management fees are guaranteed once the fund has been raised, and can be substantial. BIL took $18 million in fees in the period 1981–9 on funds of $71 million [265–6], and they were not atypical of the time, nor of VC subsequently. VCs are usually paid substantially more than other investment professionals as 'base salary': the million-dollar/pound/euro compensation packages boasted of by major financial

Table 11.1   Compensation package for top US banks

Ranking of senior employee	Base	Bonus	Total
Top	237,979	78,622	409,398
2nd	158,432	48,016	256,542
3rd	128,079	31,659	192,178
4th	113,538	23,862	165,911

Note: Composition of compensation package for partners/top executives of 'smaller' (assets < $1.4 billion) US Banks.
Source: Data from [269]. Figures in dollars.

institutions are made up primarily of bonus payments, not base salary (see, for example, [267–8]). Bankers' base salary is surprisingly small (at least, surprisingly compared to their 'headline' earnings, although still extraordinarily large compared to the scientists' base salary): Ang et al's review of the salaries paid to the top executives and partners of smaller US banks (i.e. banks with under $1.4 billion in assets under management), that is the most highly paid four or five individuals in the organisation [269] shows that even the top performing banks 'only' pay these people $0.25 million (Table 11.1).

2% 'management fee' on a $1.4 billion fund is $35 million, enough to employ 140 partners at those rates. It hardly needs to be stated that no specialist biotech VC fund has that many partners.

Thus, if European VCs are rational agents, their personal economic thinking should be dominated by their management fees, not their Carried Interest, as it is the former not the latter that will earn the majority of their reward. This means that, in reality, they have little interest in the success or failure of 'their' companies: all they have to do is spend the fund they have raised to earn the large majority of the fees.

This is exacerbated by the timing of the funding cycle, which further erodes the motivation to invest well, and in fact provides the drive behind many of the destructive practices that I have identified above.

VCs looking to raise a fund for investment in high-tech companies have a choice. They can

- raise a fund, invest it, manage the fund well for ten years, exit and at the end of ten years gain the accumulated management fees and the profit share.

Or

- raise a fund, invest it, then essentially abandon the management of the investment while they raise another fund and invest it. At the end of ten years they have the accumulated management fees of *two* funds, but minimal profit share because there is minimal profit.

Although VC theory and the VC 'sales pitch' say that they will do the first of these, the second has a much better risk: return profile. If profit share and management fees are equally valuable (which is the best case for non-US VCs in the last two decades) then running two funds in a way so as to return no profit share is economically equivalent to running one fund well, and running three funds beats them all. Therefore, once they had raised and invested one fund, rather than put continued effort into that they should ignore that fund and seek to raise another. Raising a fund takes around a year, investing it in propositions that are not clearly stupid takes another 2–3 years, and investing the fund is important: many funds pay a management fee on the basis of funds invested as well as funds committed, and if funds are not invested briskly the LPs can reasonably say that the managing partners are not doing their job and cancel the entire fund contract. So rapid investment is in the VC's interest, which will be achieved 3–4 years after the start of fund-raising. If their motivation was to raise funds as fast as possible, so as to garner management fees, we would expect VC General Partners to seek to raise a ten-year fund every 3–4 years, and this is exactly what they do (Figure 11.2).

It is reasonable to expect large investment groups such as Apax or Schroders to be running multiple funds in parallel. However, there is almost no difference between the average gap between raising funds between large groups (here defined as groups having at least $1000 million under management) and smaller, specialist biotech fund managers or smaller 'tech' funds that invest in some biotech companies: all three have an average fund gap time of around 3.5 years. Even small and specialist management groups are almost always raising their next fund before the outcome of their last fund can be objectively evaluated, and usually before more than one or two of the companies in it have come to a definitive valuation event, such as an IPO or bankruptcy. Often they are raising funds when the only thing they have done is spend their last fund, which is not a task requiring enormous commercial acumen.

This business model is clearly superior in terms of return to the VC to a business model that requires the VC to actually make a profit from their investment. However, it also raises a significant problem for the VC management group. They are asking investors to give them more money *before* the results of them investing the last lot of money have become

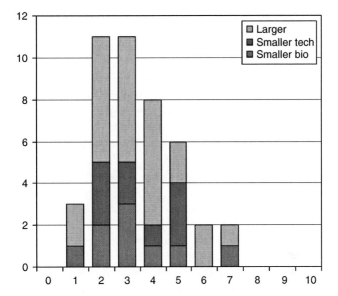

Figure 11.2 Time gap between fund closings

Note: How long between a European VC closing one fund and closing the next, for VCs that invest significantly (i.e. more than twice in the last decade) in Biotechnology companies. For the most prolific nineteen management groups whose funds have invested in biotechnology in Europe since 2000. Investment management groups included are Abingworth, Advent, Atlas, Avlar, Bankinvest, Gilde, Life Science Partners, HBM Bioventures, Healthcap, Index ventures, Merlin Biosciences, MPM Bioventures, MVM, Schroder Ventures, Scottish Enterprise Partners, Sofinnova, SV Life Sciences, Technostart, TVM. (Some other 'prolific' investors are omitted because they manage open-ended funds for which this type of analysis is not possible.)

Source: Data compiled by the author from CapitalIQ.

apparent, and in the case of a lot of European VC, when it is apparent from the public outcomes that their previous efforts have not been very successful. They are, in fact, in the same position as first-time entrepreneurs coming to them to ask for investment. So what arguments do they give?

Balboa [270] analyses the features of a VC management group that correlate with the volume of funds raised for 'developing' markets (which in US terms means anything outside the US). They find evidence for correlation of funds invested with

1. the volume of past investments
2. the ratio of portfolio companies to investment manager
3. the percentage of divestments carried out through initial public offerings and trade sales

4. the membership of the national private equity association
5. the size of funds under management

as characteristics of the highest importance in *raising* funds (i.e. convincing other investors to give them money). The first and last of these are circular – having raised and spent funds helps to raise funds. The fourth is trivial, requiring only payment of membership fees. The second we have already seen European VCs ignore on a truly spectacular scale, but in any case conflicts with the first, so presumably these two balance out when comparing funds that only invest at a specific stage in company's development. This leaves the third as being of any actual significance. This would predict that the amount that VCs raise is dependent on the number of IPOs of VC-backed companies, and this is exactly what is seen on an industry-wide scale. Historically, the variation in the number of VC-backed companies that IPO is indeed sufficient to explain almost all of the variation in how much money VCs can raise (Figure 11.3).

VC pitches are therefore aimed at providing supporting evidence that they can carry investments through to successful divestment. This is a

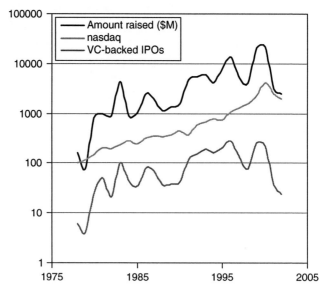

Figure 11.3   VC funds raised versus VC-backed IPOs
Note: Illustration of link between the amount of money raised by VC funds, the number of IPOs of VC-backed companies and the general stock market for tech stocks (NASDAQ composite index). X axis – date. Y axis – number of VC-backed IPOs, VC funds raised, and NASDAQ index.
Source: Data from [12], NASDAQ website.

problem for the VC if they have actually *not* achieved any exits: they must argue instead that their skills and processes are such that they can do so in the future. I have not found any systematic data on this, but I have seen a number of VC 'pitches', and there is substantial consistency. The common arguments are:

- We have experience.
- We have a selection process that identifies the top opportunities (the improbability of this needs no further amplification).
- We invest in 'hot' areas.
- We aggressively manage shareholder value.
- Look at the portfolio of our last fund so far.

The most important of these is the last, but it is worth dissecting the others to show how the need to make this pitch could drive VC behaviour.

Experience should mean experience of successful VC. Gompers and Lerner [12] comment:

> Not only is it difficult to raise as new venture capital fund without a track record, but the skills needed for successful venture capital investing are difficult and time-consuming to acquire.

We have noted that it is demonstrably not true that 'there was no VC in Europe until the 1990s' [44]. However, it is true that the large majority of European VC General Partners have little experience of VC, and those that do which have invested in early stage biotechnology usually have a record of making a loss. So the experience and success that they claim in pitching to investors is success in some other business-related area. A common one is success as an entrepreneur (even though, as we have seen, prior success as an entrepreneur is only weakly predictive of future success). In reality, VC success as an entrepreneur is weakly negatively correlated with the chance that a company that that VC invests in will achieve an IPO, and uncorrelated with whether the company will file for bankruptcy [90]. The only better predictor of VC failure is experience in consulting.[2] Predictors of success are expertise in law or finance (not accountancy, which is a different discipline entirely).

Investment in 'hot' areas means areas that others are saying is hot. It is unlikely that the generalist investment banker being addressed in a VC pitch will have the technical knowledge to decide whether (for example) population genetics or shotgun EST[3] sequencing-based strategies are the most effective target identification technologies or whether both are a load of horse feathers. This is why VC investment rockets when public markets rise, and falls when public markets fall: if the public markets are saying that genomics is 'hot', then VC piles in. It is also why VCs follow fashion to such an extraordinary degree. Logically, the decision to invest in a company *today* should not be influenced by the prospects of floating that company *today*, as you cannot float it today – it is not ready. It will be ready in 3–7 years' time (but see below), and the one thing that can be almost completely certain about the stock market is that it will not be the same in 3–7 years' time as it is today. One could even go with Winston Churchill's comment[4] about weather forecasters and invest in exactly what the stock market is *not* buying today. However, this is not an argument about business, but one about the sales pitch the VC is making today. If today's 'hot' technology on the public markets is genomics or nanotechnology or biofuels or some other technological fad, then being able to say 'we are investing in this hot area' is a good sales pitch, and if the naivety of the candidate LPs is such that they do not realise that investing in biofuels in 2007 will not necessarily be a good idea by the time the fund matures in 2017, that is their problem, not the VCs.

For the companies, however, this is a problem, as the VCs have to position their *past* investments as being in 'hot' areas. Investing in a genomics company in 1997 looked like a very clever thing to do. Today it looks a bit foolish, so why did you do it? For the pitch, the VC therefore has to make the genomics company look like what is exciting today, and as companies all have websites, conference pitches, scientific papers and other forms of publicity, this means that the appearance must be quite deep, not just a change of logo. The company must actually change its business model and operational direction, as we have noted happens. The problem for the investee company is that the rate of change of investment fashion is faster than the company can actually accommodate, and the company will spend its time, effort and cash pursuing an ever-retreating goal rather than achieving anything. So the 'what's hot' argument is actually quite destructive.

Managing shareholder value is an accurate and effective sales pitch. In most contexts, the aggressive policies that VCs pursue to increase their equity stake in a company rather than increase company value (some of which we have reviewed) would be seen as counterproductive. As Thomas Stewart commented:

> At a dinner in New York last spring [2005] a friend offered a spirited exposition of the Chicago-school argument that a corporation's only responsibility to society is to make as much money as it legally can.

That's a fine theoretical position, but the others around the table, executives all, found themselves uncomfortable. Theory (and preference) aside, you can't walk the talk. A company that shunned society would soon be ostracised in return, to its very real cost.

[271]

But VCs are not concerned with their reputation among entrepreneurs, let alone among the wider society, only with that among their potential investors, so there is no reputational downside to aggressive investment policies. That these damage the companies in ways I have reviewed in previous chapters is not an issue until later, and 'later' is not relevant as by then the funds will have been raised. For the purpose of the VC sales pitch, if you can show that the value of your shareholding has gone up, even if the value of the company has gone down, then of course that is good for *your* investors. This could be a drive behind the Pref share effects, M&A choices etc. discussed earlier.

The portfolio record so far is the last point and is the most critical. If VCs cannot show success, and most cannot because their existing funds have not matured and in any case are not that successful, they can nevertheless show *lack of failure* (i.e. the continued opportunity for future success), and the *appearance of progress*. The small size and immature experience of the European community means that its business must be based primarily on reputation, and members of that community cannot have any failures [272], so both of these aspects of the VC sales pitch are critical.

'Appearance of progress' can mean many things, but basically they are all the things listed in Chapters 7 to 10:

- driving companies towards business models that lead to the promise of huge returns in the distant future, without crystallising value to a lower number in the meantime,
- changing the business model to fit the latest trend – tying it to the current 'what is hot' fad,
- 'Directing' the business so that activity can be positioned as progress,
- M&A activity (especially if through shareholder engineering you can increase your own shareholding without investing any more cash),
- changing the management ('we have brought in a red hot new CEO') and
- move to a big building on a science park with personnel policies, ISO 9001, multiple layers of management etc. to ape larger, successful companies.

The need to appear to be backing what is hot explains two things – the heavy investment in whatever technology is currently trendy, and the drive to change business models. Even if VC partners themselves do not know that a new drug discovery technology is not likely to discover new drugs very much faster than a previous one (as we discussed, and see references [199–200]), their advisors do: unless a new technology can capture business as well as science, it is of marginal value, and the drive away from revenue-based businesses prevents the obvious routes to such technology becoming a business opportunity. Yet there are enormous waves of investment in 'the latest thing'. The reason is that 'the latest thing' looks good on the PowerPoint slides.

Avoidance of appearance of failure is, however, more important than appearance of progress because VC LPs are not stupid and they know that recruiting a new CEO is not really a substitute for having sales. However, here the VC fund structure works to their substantial advantage. Investors put money into a fund at one end and get returns out of the other. The input and output can be seen clearly. In-between the value of the fund's contents can only be estimated with the help of the VC (General Partner) interpretation: visibility of the fund's contents is almost zero for the investors in that fund. We discussed this in the context of company valuation in Chapter 3. (Contrast this to the visibility of investee companies to the VC investor, who usually requires not only a Board seat and copies of relevant financial summaries but also control over many week-to-week operational aspects of the company's management.) The VC wishes to use the image they project of the contents of the fund to raise interest from new investors to put more money into a new fund. What is *actually* going on in the fund is not as important as what is *perceived* to be going on by the outside world. The only breach of this information asymmetry is when an event occurs to raise the visibility of a company in the fund to an extent that it cannot be 'spun' by the VC: I term this a 'visibility event'.

The issue of the 'avoidance of failure' is the critical one here. Of course a rational investor would not want their company to fail, but in this case what the VC needs to do is to avoid the investee company *looking* like a failure or *looking like a potential failure*. There are two approaches to this. Firstly, make sure that the company does not do or say anything that would show it (or you as its investor) in an unfavourable light. They need to remain 'on message' in the current political argot, including remaining 'on message' to whatever the current investment pitch is in terms of business. The real 'Moral Hazard' risk between VC and entrepreneur is therefore not that entrepreneurs will act opportunistically with the funds invested in their companies. It is that they will act opportunistically with the reputation that the investor has invested in them, and so cause a negative visibility event. The money is, after all, someone else's, but the reputation is the VC's own. This is one reason behind the 'fire the founder' philosophy discussed above, as entrepreneurs are notoriously unable to stick to the corporate line, and in some cases 'shoot their mouths off' in highly uninvestable ways about the failings of their VC investor, the new CEO, the company strategy or other topics. More generally, complete control over the Board means that you stop the company doing anything that would adversely affect the image of the company you want to project. One example (which the provider asked me to keep anonymous) illustrates this.

> We wanted to spin-out the [technical division] into a new entity. There were revenues, losses, some neat technology – I knew some guys who would probably have invested. I was told that I could not do this, could not even talk to investors about the possibility of doing this, or I would be fired. It would send 'mixed messages to the invest-ment community'. In reality it would send a clear message – that the VC had messed up the opportunity and we were taking it away from them to do it properly. So the [technical division] remained where it was, did not get funding, and was closed down. Lose-lose.

Conventional economic literature states that contracts between investors and investees (Principals and Agents in Principal-Agent theory literature) are optimised to motivate the Agent to maximise return to the Principle [220, 273]. VC investment with entrepreneurs should be structured to maximise the return to the VC, that is, the value of their shares in the company, through motivating the entrepreneur. But many of the actions of VCs do not seem to be directed in this way, as we have seen. The reason is that the utility function for the VC is *not* based on the value of their equity in the investee company alone. It is based pri-marily on perceived control of the company, and the utility of the com-pany's perceived position in raising further finance.

The other, very odd consequence of this need to present an appropri-ate pitch to raise further funds is that revenue is an undesirable feature in VC-backed biotech companies. We have discussed the drive for com-panies with the potential to spend cash on research to discover drugs or to sell products and services to switch their business model to the for-mer. Why is this desirable? Because having revenue crystallises a com-pany's valuation in a way that is transparent to the Limited Partners: a revenue line appearing in the reported accounts is a visibility event for the fund, and a strongly negative one. Almost never is the revenue enough to justify a 40% annual return on the VC's investment. Consider, for example, the industry-accepted rules for valuation of a private company, discussed in Chapter 3. If a VC invested £5 million in a company, its post-money valuation must have been at least £5.5 million to make the investment make any sense. That can be talked up if they file patents, recruit a CEO and so on, but if they suddenly get a revenue stream then the revenue-based calculation takes precedence and the valuation could fall: indeed it is likely to fall, as it is unlikely that their initial revenue will be very large, and it will probably be loss-making. So the company is classified as a failure, not because sustaining a business on its own revenue is not a good thing, but because the revenue cannot sustain as good an argument for high future valuation as the speculative argument based on presumed future success in drug discovery and development can sustain. Such companies are classified as 'living dead' investments in VC terms, a blanket classification for companies that cannot achieve the promise of returns that are needed but which nevertheless will not 'die' by failing. Ruhnka et al [165] point out that in all other economic fields a company that can sustain itself on revenue is a success.

> We conclude that the concept of 'living dead' investment is unique to the context of VC investing in high-growth-potential companies, and represents a *failure of investor expectations* as distinct from an economic failure of the venture [italics in original].
>
> [165]

The sample of VCs analysed in Ruhnka's study considered that 33% of investments 'failed' in this analysis (and 39% were 'poor exit potential', which could mean the same thing – that they were economically successful but nevertheless did not perform the function that the VC required of them in the VC business model).

The flip side is the desirable nature of positive visibility events. There is only one type of any significance. VCs require exits for their investments, so that they can return money to investors (ideally more money than the investors gave them). Thus, VCs will drive for IPOs. The one aspect of the Balboa and Martí [270] study that we have not addressed is the desire for an exit event, ideally an IPO. We see that by far the most obvious determinant of how much investment institutional investors are willing to put into VC funds is the rate of IPO of VC-backed companies at that time (note that this is *all* VC-backed companies – VCs can argue that 'they made good returns doing this, and so we will too'). This is observed to be true, over time (Figure 11.3) and within any one time frame over different geographies [274]. VCs recognise this.

VCs' priority when raising new funds must therefore be to float a company (or more than one) on a wave of news if not achievement, and then maintain the wave until they can raise a new fund. For a drug discovery company this is much easier than for a company with mere profits, because there is no countervailing visibility event to balance the news flow. A new drug is found, new patents filed, a new clinical trial is started, all of these can be newsflow on which to launch a company onto the public market (Tegenero's website proudly announced that it was about to test its product TGN1412 in late 2005 – as history showed rather dramatically in this case, *starting* a clinical trial is not the same as *success* in a clinical trial [25]).

In particular, if I am a first-time VC, with a portfolio of expensive, loss-making companies and no actual track record of success, the one thing that will raise my credibility more than anything else: an IPO from my portfolio. Any guide to 'how to write a business plan' will tell the new entrepreneur that they must have a chapter about how they plan to exit, and VCs' preferred exit has always been an IPO. This has started to change in Europe, mainly because the value of IPOs has been declining and the value of trade sales of companies has been increasing, but IPO is still the preferred route, as it is far more public, and hence is a better visibility event. (Many trade sale exits of biotech companies do not state the financial terms, for example, and so the VC cannot point to any evidence of financial success.)

But companies need to develop from start-up to a sufficient size and accomplishment to be suitable for flotation. IPOs are relatively rare exit events in practice, and relatively few biotech IPOs show any post-IPO capital growth (14 out of 41 global biotech IPOs in 2003–5, according to Koberstein [275] showed positive post-IPO share price growth, the majority lost share price in a market that was overall bullish). Around 20% of VC-funded projects can cash out via IPO, with the rest being sold off, merged etc. in a mixture of modest success and 'scantily disguised failure' [276]. So IPO is unlikely, and the chance that a young portfolio will have an IPO-ready company in it is fairly small. But an IPO is the ideal way to raise the visibility of a VC fund – it is a highly positive visibility event. In the biotech industry 'it has always been an article of faith [among investors] that the number and size of the successful [IPO] exits would be substantially greater than the failures' [277]. What if the portfolio contains no companies that are suitable for flotation? Then the VC has the choice of trying to raise funds with a weaker case or pushing a company to IPO prematurely. The latter is a common choice, and is called Grandstanding.

The classic analysis of the Grandstanding effect is Gompers' 1996 study [278]. Analysing 433 IPOs, Gompers found smaller and less experienced VCs took their investee companies to the public markets earlier with less investment and less experienced VC directors, significantly underpriced, and (most significantly) at the time that the VC fund is raising new funds. The drive for this to happen is reputational – VCs need the reputational boost of a positive visibility event to help them raise funds. Lee and Wahal extend this to show that the level of underpricing of an IPO correlates with the amount of funds the VC company subsequently raises [279]: the visibility event is not merely to have an IPO, but to have one whose shares rocket, Genentech-like, after the IPO. This is simple to achieve by floating at a below-market share price, and the only downside is that the company does not gain as much cash raised and has to give away more equity to achieve it. Such early IPO and underpricing also gains the VC the reputation of providing rapid gains the value for institutional investors that invest in the IPO, which is also attractive to them [280].

Hidden in this is that the VCs themselves make *less* from an early, underpriced IPO, and therefore that their Limited Partners make less. In addition, lock-in (contractual controls on the private shareholders blocking them selling stock in the company for a period after IPO) means that they will not benefit from early gains: they will have to stay with the company until the underlying immaturity of its business (usually its technology base for a biotech company) is revealed. Again, however, this is not relevant to the VC if the aim is to gain a positive visibility event rather than to make profit: the financial gain from a positive visibility event, and consequent ability to raise new funds and garner new management fees, outweighs the reduced chance that they will gain any Carried Interest.

Several observers have commented that bringing immature companies to the market is not beneficial to the companies or the market [281]. In a systematic survey of Singaporean companies, Wang et al [282] found that VC-backed companies were taken to IPO sooner than non-VC-backed ones, and did worse in post-IPO performance. This was worse for younger VC firms than established ones, and worse for companies that had received extended VC investment rather then only mezzanine financing. (By contrast in the US, with established VCs, VC investment results in better post-IPO operational performance and long-term share price [283].) However, as we have noted, this is not of economic importance to the VC.

Perhaps the Grandstanding effect is an issue of start-up companies, not of VC itself? In the last 2–3 years a number of biotech companies have floated extremely early on AIM, in some cases such as VasTox using the public markets as their only source of significant finance. Maybe high-tech start-up companies are inherently likely to be driven to premature IPO. This is possible, but seems unlikely, because another source of equity investment into early stage private companies which does not demonstrate these Grandstanding effects. Corporate venture funds have similar fractions of investments that complete IPO, but at 50% higher valuations, and fewer liquidations than VC groups, and the patterns are not related to technology, or to the 'fit' of the technology with the parent company [12].

Interestingly, they also do not have the prolific investment pattern of VCs, investing at a rate of 2.5 investments per year versus 7.1 for VC funds. We have noted above that the combination of VC's seizure of management control with their overload of commitment results in poor management capabilities. It is likely that corporate venture investee companies perform better at least in part because of this lower management overload, and hence improved quality of input into the investee companies. For them, the story of 'investor added value' might actually be a reality.

The analysis above implies that VCs seek, and gain, investment in new funds before any measurable outcomes from old funds are available, and do so through actions that are likely to damage the value of those previous funds. It seems extraordinary that investors in VC funds – primarily pension funds, large equity and mutual funds, national and transnational banks and similar institutions, should invest in this business. But they do, and furthermore they do not appear worried by the obvious lack of results that VCs have presented them with: the 'pitch' outlined above is enough to convince them to invest. As most of the criteria for judging a VC fund are dependent on activity of raising and spending investors cash, that would imply that a VC that has received and spent a lot of money should appear more successful than one that has not raised so much money, *regardless* of how well they have invested it. This extraordinary idea (after all, how hard is it to get rid of £100 million?) is supported by the apparently inexorable increase in the size of funds that VC management groups raise, each larger than the one before, supported by the 'evidence' of their success (Figure 11.4) and the complete lack of correlation between the increase in fund size fund-to-fund and the time between them (Figure 11.5).

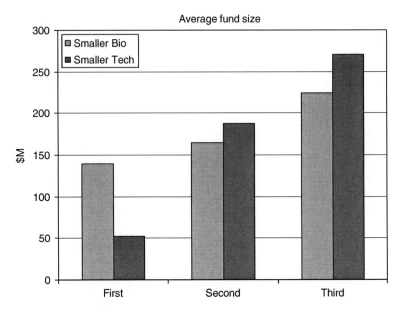

Figure 11.4   Average size of specialist VC funds in Bio- and general tech
Source: Data from the same VC set as Figure 11.2.

Note that in Figure 11.5 some managers *do* raise smaller funds than their 'last one'. Nearly all of these were fund-raisings when the first fund was raised in 2000–1, and the second in 2004–5, respectively the zenith and nadir of fund-raising in European VC.

The real business model for VC is therefore clear. The VC seeks to raise funds in parallel, multiplying the reliable revenue stream from management fees at the expense of developing the value of their investee companies, and hence of profit share. The reputational features of raising and spending each fund are used to raise the next, bigger fund. The commercial outcomes from the investments are unimportant. The critical goal of a VC must be to preserve the value of their fund as a PR pitch, by driving for exits if they can, maintaining an appearance of progress and keeping investee companies from pitch-breaking visibility events if they cannot.

The fallout of this is the need for tight control of companies to prevent them creating adverse visibility events for the VC's own target market. This leads to replacement of noisy and often obstreperous entrepreneurs with politically aware 'professional' management, drive towards business model that promises enormous returns in the future while avoiding them today, following the trend in business models to achieve this and

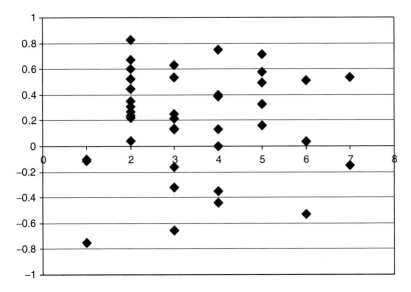

Figure 11.5   Fund increase versus inter-fund gap
Note: X axis – time gap between raising funds. Y axis – increase in size of fund, plotted as $(F_1-F_2)/(F_1+F_2)$, where F_1 and F_2 are the sizes of the two funds. R^2 for linear equation fit = 0.002.
Source: Data from the same VC set as Figure 11.2.

to be able to say that they are 'in' whatever the public markets believe is a 'hot' business, driving companies to behaviour that mimics larger, successful companies, and seizing larger shares of equity even if this damages the company. Above all, they need to keep tight control of all company actions, not to make them do things but to prevent them from doing things.

Thus, the apparently bizarre and counter-productive behaviour we saw in previous chapters is seen in this light as the rational implementation of a business that makes VCs very good money indeed.

This model also explains why the creation of a large number of 'academic' spin-out companies, of the sort noted in Chapter 3 above, is a bad thing. VCs require a portfolio of companies that, two to four years after their investment, will look good for their future fund-raising. One way a company can look good is if its science is credited with being world-class, and of course is in the current 'hot' area. Newly created companies based on academic science that is at the forefront of knowledge and technology will of course be excellent examples of such companies. Failings in their business model, that is, their ability to turn that science and technology into commercial end points, are unlikely to be evident for as long as the initial investors can keep them going. In fact, failure of the science itself is not important, and the risk of scientific failure is not analysed in any serious depth as we saw in Chapter 5, providing it does not happen too soon.

The wave of state-supported spin-outs in Europe have been very helpful to the VC business model in this regard. The science is undoubtedly world-class (as discussed in Chapter 4), and the founders were often inexperienced and naive, and hence would do what the VCs tell them to do. The founders are also not expecting to be CEO (at least in part because the VC will tell them that they are incompetent to be CEO, and they believe them), and so are amenable to management change, and will accept other changes such as moves to Science Parks, employment of expensive financial and HR advisors, and other features that we have discussed. The quality of the science (as opposed to the products which are promised to emerge from the science) is likely to be excellent, and hence not to hit any very public roadblocks in the 2–3 years after investment, especially if only a small amount is invested so the science cannot be taken forward very fast or adventurously. The only problem is that the business is not sustainable, but this need not be apparent until 2–3 years after investment, at which time the next fund has been raised. It is not coincidental that the calls for changes in the way that university-spin-outs are encouraged and funded came 3–4 years after the enthusiasm for their creation in both Germany and the UK – this is the timescale for VC investors to invest in them, use that investment as leverage to raise a new fund, and then abandon them.

This is not much fun for the companies concerned, but realistically if they never had a business that could work they should consider themselves lucky they got any investment. However, the VC business model means that it is bad for the industry as a whole. We have seen that VCs do not have a motivation to distinguish between 'real companies' and these grant-supported start-ups. In Chapter 4 I argued that the failure of European biotechnology was not the result of lack of 'real companies', but now we can see that this was not the whole story – if there are a lot of 'unreal' companies receiving VC funding simply because they look good in the pitch used to raise the next fund, this *will* deplete the VC pool of funds to support good companies. It is true that there are good propositions, but the excess of bad propositions which nevertheless get funded means that the good propositions will not get funded, or not get funded to an appropriate level.[5] These programmes of company creation are therefore not neutral. They are damaging, not because they *should* be according to rational economic models, but because the VCs are not acting on the rational model that their PR says that they follow.

The one aspect of VC behaviour that is not well explained by this model of their economic motivation is why they invest so little in each opportunity. If the VC business model is truly agnostic to eventual investee success, then it should not matter whether they invested a little or a lot. So they might as well invest a lot, to gain the benefit from the improbable eventuality that their investments do generate enough profit to give them some Carried Interest (profit share) from the fund. It is plausible to suggest three reasons for this chronic underinvestment: the true motivation might be a combination of these, or of these and other factors.

Firstly, that it is what everyone does in Europe. If reputation is the critical factor that determines the VC management group's financial success, running with the pack will be an important part of the pitch that confirms their reputation without substantial effort. This is a variation on the 'we invest in hot areas' argument, and is in fact the argument that many VCs give to potential investee companies when discussing valuation, along the lines of 'No-one values a Series A at over £2 million nowadays'. This argument makes no sense in terms of the actual value of a company, but makes excellent sense in terms of VC reputation and positioning with its future investors. This is related to the control that VCs demand in order to avoid negative visibility events. If a company is given 'too much' money, then it is hard to control what it does – it might, for example, stick to a business model long after it has ceased to become a 'hot' area: if the company is perpetually about to run out of money, then its management is unable to achieve anything, even execute the current business plan, without VC acquiescence.

Secondly, if a fund invests a small amount of money in a large number of companies, then if one of those fails spectacularly (i.e. achieves an objectively verifiable negative visibility event) its impact on the fund overall will be small, and can be 'spun' in future fund-raising. If a fund makes only a small number of investments, then the failure of one is going to have a clear impact on the performance of the fund overall, and this will be obvious to future LPs. Of course, investing effective amounts in the companies will increase their chance of substantial success, but this will occur later (in the biotechnology industry case), and so the counterbalancing positive visibility events that effective investment levels would bring are not useful in raising new funds – they come too late.

Lastly, there is a simple economic argument. Each investment earns the VC management group funds, in terms of directors' fees and consultancy costs. An investor director for a UK biotech start-up is typically paid between £10,000 and £15,000 per annum – in some investment groups this goes to the management company (as it should, as acting as a director is part of the management group's function), in some it goes to the investor director personally. In addition, some early stage VC investors 'strongly suggest' that investee companies execute contracts with the VC management group to provide consultancy support, to provide office and administrative facilities, or to provide financial control services for the company. (Recall that in Chapter 7 we discussed that the theory of VC, and apparently the practice in the US, says that this added support for companies was part of the value of VC investment itself, and the reason that VC money was expensive.) Contracts range from £10,000 to £50,000 per annum. Thus, a VC investor can expect additional income, some of it directly to them, of between £10,000 and £65,000 per annum per investment. If the size of a fund is fixed, the economic argument is therefore to invest as little as possible into as many investments as possible to maximise this income stream. This is supported by the extraordinary number of Board seats that some investor directors have.

It is not clear if these arguments, in combination, are enough to offset the undisputed negative effects of investing too little in investee companies, and hence making almost certain that the VC fund managers will have no profit share. Individually they seem weak, but a combination of fear of failure and of breaking ranks and significant personal financial reward may be sufficient to explain the systematic underinvestment in early stage companies.

12
What to Do about It: Government

I have reached the point where I am going to try to suggest some solutions to the problems outlined above. As I warned at the start, none of these are complete or enormously compelling. Complex problems rarely have simple, dramatic solutions: they yield to incremental improvement on a number of fronts. This is the approach I advocate.

I would also emphasise that not all VCs run the business model outlined in Chapter 11. Some, even in Europe, aim to invest well, manage well and make money through realising their shareholding in valuable companies. It would be invidious to name specific companies, but a few seem willing to be part of a team with start-ups and build value together. Entrepreneurs need to seek them out. For the rest, the following are some ideas about how to encourage more constructive behaviour from them.

The obvious answer is to ask VCs to do better. As VC partners in a specialist biotech fund management company earn mid six-figure sums (pounds) from the current business model, and senior partners are into seven figures, there does not seem much motivation for them to do that. The business model I outlined above is consistent, successful and apparently sustainable. While it operates there are strong economic incentives that result in a small, underperforming European biotechnology industry. So we should instead ask if there are rational ways that the environment in which this aberrant economy flourishes can be changed, and changing the economic environment is generally considered a task for governments.

Governments think that biotechnology companies are a 'good thing'. They are clean, require little investment in manufacturing infrastructure, exploit the knowledge that their universities are believed to generate and have an international presence to raise the country's profile.

Most biotech companies are small, and the economics literature generally concludes that long-term economic growth comes from small, nimble, innovative companies. They capitalise on investment in scientific excellence. Governments therefore want to encourage the formation of biotechnology companies.

In Chapter 1 I used the term Cargo Cult Economics to describe this logic, and I do not think it is too strong a phrase. A small number of small biotech companies have grown to a large size and to profitability. Therefore, if we set up a small biotech company it will grow to be large and profitable. The invalidity of this argument, stated in such a stripped down form, hardly needs to be demonstrated, especially in light of the historical failure rate of biotech companies. But there is a continuing belief that *new companies* are the answer. This results in a huge feedstock of poor companies that are only too grateful to take the VC dollar or euro, and willing to fit in with the requirements to change management, change business model, move to upmarket premises etc. because of their naivety and lack of any other choice, especially once VC has invested and imposed the control over the company described in Chapter 7. Government initiatives thus provide feedstock not for the theoretical VC business model, which would not invest in them, but for the actual VC business as laid out in Chapter 11.

So we can ask two questions. Why should governments help biotech companies at all? And why should governments support the biotech investment industry, as it is presently constituted?

It is widely believed that this exploitation of the science base, and particularly the creation of new, venture-backed enterprises developing intellectual property licensed from academia, benefits the national economy in which it occurs [12, 42], although whether it really does is a matter of debate [284–6] to which we will return. The UK Government alone has published 14 reports into commercialising high tech R&D between 1993 and 2006: all assumed that this was a Good Thing. Much of the debate around government support for biotech is tied up in discussions of financing. But is biotech *per se* a worthwhile thing for governments to support?

The arguments for state support of any high-tech industry, and specifically for biotech, fall into three general camps:

1. It is worthwhile for society as a whole.
2. It is important for future economic growth.
3. It is a prudent additional investment to realise the prior investment in academic R&D.

The first of these is a statement about societal goals, and one that does not fall within the scope of this book. If society believes that biotechnology companies are a Good Thing, either because they are the only mechanism that can create products that society wants or because their activities are inherently good for their own sake, then society should act on that belief.

But why should they believe that? The application of these arguments to demonstrate why the government should support the biotechnology *industry* (and its investment) cannot lie on the foundations of furthering knowledge for its own sake – so-called pure research – as Universities are usually better at doing that, and that is (or was) their role. So it must lie in the application of that knowledge. There is, for example, a strong school of thought that drugs are only developed effectively in a commercial environment, and so commercial entities to develop new medicines must be supported.

However, in terms of *social benefit,* this argument does not explain why a government should support biotechnology companies in its own country. It would make equal sense for a European government to support biotechnology companies in the US, and have their pharmacies or state health system buy drugs from those companies (as indeed they do). The argument also does not provide reasons why these entities should be new biotech companies, as opposed to established ones or indeed conventional drug companies. The same argument applies for any technological area, not just health care. A 'wind farm' clearly has to be in the windy parts of the country and as near to the consumers of electricity as is practical, but companies developing technology, components, control systems or other bits that go into wind farms can be anywhere. There is no *social* reason not to support wind farm technology in (say) Holland or Tierra del Fuego and import it into the UK or the US – the end result, apart from local economic concerns, is the same.

In reality, nearly all arguments 'for the good of society', when challenged with the thought that technology can be developed half a world away from where it is implemented, become arguments for the good of our society *and economy*. We want this 'social good' to generate wealth at home as well as social benefit such as longevity, reduced carbon emissions and so on.

This leads us to the second and third arguments. These at least are internally consistent, but they have some significant flaws when compared to the facts of how economies grow, which I will discuss briefly before moving on to the specifics of state support for investment in biotech.

Table 12.1 Company change as a result of innovation

Level of innovation	In what	FTEs	Turnover	Productivity	Margins
Manufacturing					
Novel	Process	0.02	−0.105	−0.07	−0.072
	Product	0.211	**−0.176**	**−1.89**	−0.19
Incremental	Process	0.002	−0.002	0.013	0.002
	Product	0.025	−0.166	**−1.83**	0.071
Service					
Novel	Process	0.02	0.064	0.035	−0.034
	Product	**0.194**	0.044	−0.059	−0.041
Incremental	process	0.034	**0.105**	**0.135**	0.037
	product	−0.058	−0.053	−0.075	−0.038

Note: Beta values for change resulting from innovation, categorised as 'novel' (i.e. companies doing something substantially new) or incremental (i.e. companies modifying and improving what they already do). Values in bold are significant at the 95% level.
Source: Data from [60].

Governments like innovative, high-tech companies because they believe that such companies have a substantial impact on employment. Generally, governments have limited interest in such abstracts as GDP unless they are reflected by reduction in unemployment figures and consequent increases in taxation and decrease in the number of unemployed voters.

Freel [60] studied the effects of different types of innovation on company performance: these results are summarised in Table 12.1.

For companies developing new products, the impact of innovation on the number of Full-time equivalents (FTEs) employed is generally positive, the effect on turnover, productivity and margins generally negative, which certainly matches with the biotechnology industry's ability to hire people and lose money. The only clear pattern of *significant* impact though is that innovation for product companies generally affects only turnover and productivity, which it reduces, and for service companies incremental improvement in their processes improves business performance. As we have noted, VC investors rarely invest in service companies in biotechnology, so this effect is of limited relevance. So objectively this data suggests that there is no point from a national economic point of view of encouraging existing, established biotechnology companies.

Studies such as Freel's focus on established companies, but it is widely believed that true innovation comes not from existing companies that change their practices but from new businesses, and particularly new start-ups. So the ideal for government is seen to support new start-ups developing innovative new products. This, however, is more of an ideological argument than an economic one, even in the case of employment. Storey and Johnson [287] analyse the argument that small company creation builds employment for the case of the UK, one of the first European countries to apply this argument extensively. Following mainly ideological arguments, the UK government in 1986 stated that the way to decrease unemployment was to create more small businesses and more self-employment. Hence they promoted 'enterprise' and job creation in small firms. This argument was in turn based on the Birch study that said two-third of new jobs created in US in 1969 to 1976 were in companies employing less than 20 people [288]. Similar studies have been done in the UK. The data itself is quite contentious, as it has to be manipulated substantially to allow for incompleteness of the data sets, and small changes in assumptions about the manipulations can substantially alter the figures. Highly successful small firms are very visible, but when small firms lose people they go bust, and so vanish rather then being still in the records for comparison. Depending on how you correct this you get 70% of new jobs or 40% being created by new firms.

But anecdotally, huge new industries are created out of small start-up firms – again, we cite Genentech, Amgen, Google, Microsoft, Dell, Body Shop, even Woolworth's in a previous generation. Partly as a result of this anecdotal evidence, and partly because of the policy studies cited above and others, there was a trend in 1980s policy circles to believe that start-ups were the best way to commercialise anything, and in particular in high-tech product opportunities deriving from university IP, was through new companies (variously called start-ups or spin-outs depending on taste). Whatever the arguments about data manipulation, the approach works in that international comparisons show a negative correlation between unemployment and the fraction employed in small firms.

But in fact there is no such correlation: even using the full data set listed by [287], there is only a -0.2 correlation coefficient between unemployment and the fraction of people employed in small companies. If Japan is omitted, on the basis that it has a substantially different social structure from all the other countries and consequently the economic drivers behind the constructs of 'small company' and 'unemployed' are different, then the correlation is almost exactly zero (+0.06) (Figure 12.1).

Other research in the 1990s also suggested that driving the creation of new, small companies, and specifically biotechnology companies, was not the best way to create economic growth [26–8, 80, 285, 289], and we have seen that, in drug discovery (the area of biotechnology most favoured by VC investors) 'small' companies are definitely not the way

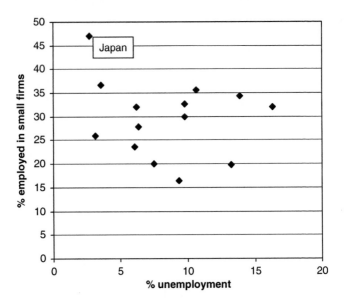

Figure 12.1   Unemployment versus small firm employment
Note: For US, UK, Germany, Finland, Sweden, France, Australia, Italy, Netherlands, Ireland, Belgium, Denmark, Austria and Japan.
Source: Data from [287].

to create success: success derives from scale. A round table discussion among US university technology transfer officers in 2004 [284] suggested that less than 1% of potentially commercial ideas warranted the investment needed to create a successful start-up company, as opposed to licensing or selling the technology to an existing body. Deakins [80] points out that the rise of very small companies (less that ten people) in the UK since the 1980s is at best only partly related to economic growth. Rather it is an effect of large-scale downsizing in large companies, and in part due to changes in tax and corporate laws which mean that businesses operating previously as 'sole traders' (i.e. non-incorporated individuals and small groups) have had motivation to incorporate and so be counted in the figures.

However, despite the lack of any evidence that creating small companies actually helps the national employment statistics, there has been a tide of initiatives to 'help' new, small company creation between: over 100 between 1979 and 1987 in the UK, and the change of government in 1987 did not seem to stem this tide: Ross [290] lists two training and education schemes, seven grant schemes, four mentoring/advice schemes and an investment scheme operating in Scotland alone in late 1990s, as well as national and EU schemes that companies could apply for, if they had the stamina. By 2003 government policy was starting to question the obsession with start-ups [286], but there is little evidence in 2007 of this resulting in changes in priorities or funding structures.

So why does encouraging these firms' creation not help the economy? Storey and Johnson's analysis is that helping new firms to be created and to survive does not help *good* firms. It just increases the numbers. They conclude that there is no shortage of small firms; the shortage is of *quality* companies. Sponsoring company creation just creates unfair competition, driving out quality firms with dross. We have mentioned this effect in investment in biotechnology across Europe: while in theory the creation of many mediocre businesses should not affect VC investment in good ones, in reality it does for reasons discussed in Chapter 11. Storey and Johnson find this to be a general effect.

The problem is that the barrier to setting up a *company*, and hence being eligible for the various support initiatives that support companies, is very low. Even before the internet a company could be set up for a few hundred pounds in the UK, and it now costs around £25: acquiring the web domain might cost another £5, so for £30 you have a corporate entity with a globe-spanning presence! This is valuable progress providing a company is seen as a legal vehicle to conduct a business and not an end in itself. However, if the formation of companies is seen as a valuable end, then government initiatives will result in the creation of companies. This is what is seen: Figure 12.2 shows the formation of biotechnology companies in three large European economies in the last ten years. If company formation was related in any way to economic conditions or scientific breakthroughs, then the timing of waves of company creation would be at least similar in each country. In fact, they are almost completely unrelated (Figure 12.2).

Instead of being related to the European or global economy, they are related to the corresponding state initiatives aimed at stimulating the creation of biotechnology companies in their respective countries: Biotechnology means Business (1994) and the University Challenge Fund (UCF) initiative (2000–1) in the UK, the BioRegio initiatives (1995–6) in Germany and the Innovation Act in France (1999). (The timing is not exact in part because of the delay in getting these programmes implemented, and actually extracting the cash from the responsible government departments.) These initiatives have therefore been extremely successful in creating *companies* to take advantage of the grants available, but signally ineffective at creating *businesses*.

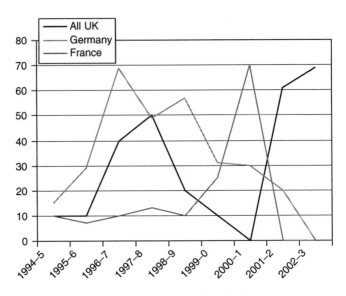

Figure 12.2   Company creation versus national initiatives
Note: In UK, France and Germany.
Source: Data from [26–8]

Germany is the clearest example of this phenomenon, because the number of small and mid-sized biotechnology companies there, was relatively low before 1995. Between 1995 and 2000 a large number of German biotechnology companies were created. They could not all succeed because there was not enough capital available to support them all through to IPO and beyond. This problem was widely understood by 1997 [291–2], but the policy of encouraging company formation is nevertheless pursued in various less florid forms to this day: in 2004–5 a new German programme was initiated to provide 500 million euro matching funds to VCs to fill 'equity gap' by experienced German biotech companies [111]. Naturally, the companies and the VCs did not object to this.

The inappropriateness of this approach is illustrated also by the number of US companies that took the opportunity offered by the German largesse to set up German subsidiaries: for example, Artemis bought up and enlarged Exelesis, Ribozyme Pharmaceuticals created Atugen [293]. Under no reasonable interpretation could this be said to be creating anything other than temporary employment in Germany.

All these schemes, in a variety of forms of words, stated that they only support companies with a clear business proposal, so they are not just causing companies to be created without creating a business for them to execute. In practice, this is a weakly applied criterion. Turok, studying Local Enterprise Committee bids for EDRF support to help the local business infrastructure for companies in Scotland in the 1990s, found that over 50% of applications did not make it clear why they were applying. Most stated some version of 'this will make the project bigger' [294]. Three bids were from Scottish agencies who would then distribute it to other agencies who would use it. All applications were, in theory, required to show evidence for demand for the service, and evidence that the service was distinctive and complimentary to existing services. In practice, only around half showed any distinction or complementarity, and under a quarter showed evidence for a demand for what they were providing.

The comparison can be made with other approaches, and particularly with the Department of Trade and Industry's (DTI) SMART award scheme and the US Small Business Innovation Research (SBIR) programme. The SMART award scheme was aimed at *projects* almost regardless of the state of the company. It matched the SBIR scheme closely (although, typically for the UK, the scheme gave grants of up to £250k rather than $25 million), in funding pre-competitive applied R&D aimed at a specific business opportunity. Many of the early, successful start-ups in the UK have stated publicly that they owe their survival and subsequent success to the SMART scheme. This scheme was closed down in 2000 and replaced by a complex system of grants that provides much more restricted support.

The SBIR scheme is widely seen as a model of how to support pre-competitive R&D that is of strategic interest to US industry. Affymetrix, for example, gained eight-figure sums from this programme to support its dominance of the world's gene chip technology, even though at the time it was too large to have qualified for UK-style SME grants. Again, this programme aims at *business projects* rather than companies. Rather than the Cargo Cult Economics of creating companies because companies are sometimes the home of successful businesses, the aim is to create successful businesses.

However, there is another, more fundamental idea behind this support for small companies that is not supported by historical observation, at least in biotechnology, and that is that governments *can* influence the growth of industry. Walsh observed that the US had no official 'biotechnology strategy' before or during the 1980s company formation boom – no 'ministry of industry' or DTI equivalent, and certainly no state or national biotechnology strategy [47]. What it did provide was major funds for specific things that it wanted to achieve – from the Department of Defence (for genome sequencing), and especially from the National Institutes of Health for a variety of health care-related R&D programmes. The leading European biotechnology companies – Serono, Novo and in their day British Biotechnology and Celltech – were set up before national or European governments decided that the industry needed their help. In fact, one could say that the only significant successes in European biotechnology happened before government started to form policy and 'help' the industry, although this would be to overemphasise the effect that almost countless state initiatives and programmes have had.

So we can conclude that the only useful way for the state to help the development of a sustainable biotechnology industry, as opposed to providing temporary employment through what is in essence state subsidy of uneconomic jobs, is to provide support for the development of new *business*. The overhaul of the cats cradle of current programmes to achieve this end would both remove the drive to found companies that sustain the business model described in Chapter 11, but also, as a welcome side effect, make it simpler for small companies without the resources of full-time 'grant officers' to identify programmes that might be appropriate for them.

That is not to say that governments are impotent. They can certainly *block* the development of an industry or an economic sector. An example is UK tax changes in 2003, which made founder shares in spin-outs derived from a University employee's work at the University a taxable benefit. This was claimed to have had the effect of stopping UK spin-out formation for 18 months [295–6] although in reality the complete lack of seed- and Series-A round financing in the previous two years had done this already. (It is also not unreasonable for employees who have been given a huge asset as a direct result of their paid employment to be asked to pay tax on it like any other benefit, although asking them to pay for it when it is not liquid is harsh, especially as subsequent VC action is likely to make it worthless.) In fact, most entrepreneurs found a mechanism though convertible preference shares to get round this [295], but this made share structures even more complicated, to the detriment of company management and motivation.

Governments can also do a limited amount to help industry, by tackling very specific issues. In the UK the rise of animal 'rights' terrorism has deterred many investors and entrepreneurs from the life sciences field, and some specific government intervention to reassert the rights of individuals and companies pursuing lawful business has helped to redress this. (The Labour government then counteracted this in 2006 with a revision of anti-discrimination legislation that defines 'belief' so broadly that it includes philosopho-political belief systems such as animal 'rights' extremism: this makes it illegal to discriminate against such extreme activists on the basis of their beliefs, even when interviewing them for employment in a company conducting animal experiments.) However, this is addressing a specific problem, rather than planning overarching strategies.

Despite the rather unimpressive evidence that government intervention can help grow genuine businesses, there is one area of state support and intervention which has been strongly supported by academia, entrepreneurs and investors. This area is government support for investment in biotechnology.

Governments across Europe have provided many support mechanisms for investment in biotechnology, usually as part of a wider programme to support investment in new, 'knowledge economy' companies. Such support ranges from tax breaks on angel investors and early investors in any small company to state investment in, or tax breaks for, funds targeted at high-tech start-up companies. Many companies would not be around today were it not for this investment support.[1] Most European governments have initiatives of this sort. The arguments for it usually derive from the existence of a funding 'gap' between what academic research produces and what commercial investors will invest in. Governments provide a bridge over this gap by coinvesting with investors who support very early stage companies, providing funds to create new investment vehicles that invest solely in very early stage companies, or investing in commercial vehicles on advantageous terms.

Tax breaks for investors who invest in early stage companies are a common concept, and widely practiced in Europe. In the UK the Venture Capital Trusts (VCTs) are a typical example [297], set up to invest in early stage private companies and companies listed on the UK AIM market (which specialises in small companies), and providing tax breaks to their investors. The UK's VCTs parallel similar schemes in the rest of Europe, Canada, the US and elsewhere. The evidence suggests that they are not very efficient investors [297], but that is probably the least of the problems with this model, as we shall discuss.

Government schemes that invest in companies or in other funds abound. Examples in the UK are funds run by, or through the auspices of, many of the Regional Development Agencies. This is not a UK-only phenomenon: the European Investment Fund (EIF) is a major investor in funds that themselves invest in companies, Israel's Yozma funds are

Government backed funds to invest directly in companies (20%) and in VC funds (80%) [53].

Examples of state-funded funds which are (at least in theory) independent investment vehicles are the UK UCFs, such as Lachesis, SULIS, Cascade, Isis Innovations and others. The UK Government has been credited, through creation of such mechanisms, with 'improving the funding environment' for biotech [116]. The funds are quite explicitly meant to help companies attract further funding: they are meant to seed companies for future investment, rather than to be an investment vehicle that can take companies from foundation to sustainability [298]. Their success as defined by their original brief [299] was

> to increase the number of good ideas, originating in universities, which are developed to the stage that they warrant and are able to attract funding through existing channels. One measure of the University Challenge Fund's success will therefore be *whether it leads to a significant increase in the number of 'deals' in which venture capital funds*, business angels, etc. decide to invest and/or to a larger number of licences with existing companies [emphasis added].

And we can note that no other *specific* criterion for success was claimed, although the general wish to 'make the nation more innovative' was stated. This focus on finance was continued for some time – see the comment by the then Minister for Trade and Industry in 2003 [300], which defines 'good' investee companies as ones 'raising further investment'.

There are two fallacies behind the argument that this is a good thing to do, and a worthwhile use of taxpayers' money. The first is that there is no 'equity gap' that can be filled with state funds. The reason companies – good or otherwise – cannot obtain VC investment is not because of a 'gap' but a wholesale unwillingness to invest, despite their rhetoric. The second is that VCs do not actually support the goals that government believe they support, of generating and building new industry and new business.

The 'equity gap' fallacy should be evident from our previous discussion of the equity gap in Chapter 5. There is not a 'gap' in the equity market in European biotechnology: there is a canyon, labelled 'venture capital'. The reluctance of VC to invest in a new company may be related to its being 'too early', but investing in those companies through non-VC mechanisms will not help. As the Chemistry Leadership Council noted [112], all that the UCFs had done was to move the 'equity gap' to later stages, so that, instead of turning down penniless university spin-outs for adequate funding, VCs now refuse to give adequate funding to more mature companies.

If the UCF were able to invest in companies that did not need subsequent investment then this would not be an issue. They would help create new companies, the companies would grow under their care and then be sustainable entities. Some UCF-funded companies do this. However, the large majority of biotech companies are set up under the assumption that they will receive further, VC funding, and as we have seen the UCF's fundamental rationale assumes that this will be the case.

This incorrect belief that state-supported investment mechanisms can help new companies grow is not a UK, or even European, problem. Johnstone, commenting on the attempts by government to fill a proposed 'equity gap' in Cape Breton (Nova Scotia), commented [301]:

> Over the past 35 years a litany of government schemes aimed at redeveloping the economy have been attempted. These schemes have not reduced the high rate of unemployment.

Coinvestment schemes such as the EIF and the recently initiated Enterprise Challenge Funds (the successor to the UCF), where state support is channeled into quasi-commercial entities that coinvest with more commercial VCs, are also not solutions to the 'equity gap', even if one existed.

But the government support for a VC funding model would be flawed even if it did encourage VC investment, because in Europe VC does not support the state aims of growing business, employment and wealth. European VCs had little impact on company growth or job creation, as we have seen in previous chapters and as others have also found [302]. The EIF have looked at all those employed by companies funded by funds they have invested in, and claim to have induced innovation, including 4400 new jobs (at a cost to EIF alone of nearly 27,000 euro per job), and of the 290 investments made by the funds they invested in, 225 were in start-up or seed companies [23]. As they only invest in funds that say they will invest in those companies, that is not surprising, and it asks nothing about whether other investors are putting in enough cash to build real businesses, or whether they are just spending EIF's money and taking the management fees. Yozma funds [53] are similarly credited with being a major force in driving the creation of the Israeli VC-backed industry. We have commented before on similar claims from UK VCs (e.g. [78]). While each claim is hard to refute, they cannot all be right: the biotechnology industry simply does not have that many jobs, and is not growing all that spectacularly.

What these measures have actually done is bolster the *VC* 's business model, by providing them with cash-poor companies that have become locked into continued requirements for investment by their seed stage funders. They also provide funds, to pay for the management fees (the largest single grant from the UK's Biotechnology Means Business initiative in the mid 1990s was claimed to have been made to a VC management company). The statistics show that government intervention simply has not overcome the 'equity gap', and previous chapters explain why.

Out of this not very encouraging picture we can identify some things that governments could do to revive Europe's biotechnology industry. The Politician's Syllogism means that 'leave well alone' cannot be an option for governments in the twenty-first century.[2] There are three areas of action that European, and particularly the UK, government could do to make the worst excesses of VC opportunism less attractive.

The first is to realign state support for innovative projects to support *business* rather than *companies*. If state support is needed to assist innovation (and this is itself a debate that has not been held for some time), then it should be focused on projects that lead to desirable economic end points, that is, new business opportunities. Two examples of such programmes which worked extremely well in catalysing the creation of innovative businesses in their respective countries exist in the US SBIR programme and the UK SMART programme. A particular value of these programmes is that they are not specific to a model of the funding cycle, and so they are neither trapped into serving the VC investment business, nor likely to become redundant as soon as the investment climate changes. Such programmes should replace the plethora of schemes to help start-up companies *as companies* being touted in Europe today. The announcement in March 2007 [304] that the UK government planned to provide much expanded state funding for coinvestment in clinical trials could provide such support providing that it funds clinical trials *per se*, and not specifically at the financing of SME companies that conduct (or promise to conduct) clinical trials[3].

The second, and related to this, is to reduce support for the VC business itself. The rationale behind the state supporting investment schemes is doubly flawed. It is argued that investment in VC funds is a more efficient way of supporting new, high-tech companies than direct subsidy. This was the rationale behind setting up the EIF [305]. However, this assumes that investment in VC drives effective investment in start-up companies – most of this book demonstrates that this is not so. What it does do is support the VC's own business model. VCs can point to government support for their funds, either through direct investment in the fund or coinvestment in companies, as evidence for the validity of their own business model, which, as I have laid out in Chapter 11, does not rely on success of investee companies to make money. Therefore not only does government support for VC investment not achieve the benefit it is meant to, it supports the VC business model that does the opposite.

Government support for *investment* in start-up, high-tech *companies* should be substantially reduced, in favour of support for *business projects*. Direct support of commercial funds should, in this author's opinion, be stopped altogether. Commercial funds are either commercial, in which case they do not need state support, or they are not, in which case their support is a grant and should be treated as such. Unlike high-tech product development, VC fund-raising and investment requires no investment in R&D, no new technical or product concepts, and (as we have seen) limited risk for the management group operating the fund. The only rationale for supporting such activity is that it supports the companies in which it invests – that being so, the state (if it chooses to support such activity) should support those companies directly and cut out the well-paid middleman.

There are some issues around EU state support rules which come into play here. The EU has rules that prevent governments from propping up failing companies, which are designed to prevent governments from keeping their national airline and steel industries from genuine commercial competition. The same rules also mean that SMEs are generally not eligible for grants to do research. But if the grants are called investment, that is OK, if some other hoops are jumped through. The US has no such limitations on its support for long-term, strategic technologies. The source of this apparent technophobia in Europe is beyond this book, but it does mean that what is in essence grant support can look very strangely delivered to the outsider.

The third is to tighten up regulation on aspects of the operation of private equity investors in private companies, to make less attractive those activities which support their current, unproductive business models. Among these would be tightening up accounting standards so that VC firms have to report more realistically on the value of their portfolio (c.f. the discussion on the *actual* portfolio values seen in Chapter 3), and making more transparent the fraction of invested funds that are spent on non-profit-related management fees.

A specific legal change would be to make those who act as directors of the company *de facto* acquire *de jure* responsibilities of directors.

As I noted in Chapter 7, private investors effectively control the investee company. Non-director investors therefore have the power of directors but none of the legal responsibility: conversely non-investor directors have all the responsibility but none of the power. Non-investor directors, it can be argued, can always resign, but if they are not told what is going on (and this happens in such circumstances), then they will not know that they should do so for their own protection until it is too late. At the very least Company law should make it a requirement to publish any contract which limits directors' freedom of action, such as shareholder agreements imposing substantial controls on directors' freedom of action, so that a defence of powerlessness can be presented more easily, and so that the real controllers of the company can be seen if potentially unlawful acts are being executed.

One such class of act is the oppression of the rights of minority shareholders, so the last item in this discussion of the law should be minority shareholder protection. There is no doubt that minority shareholders are oppressed in VC-funded companies – even if the company does well, their shareholding can fall in value or be deleted from the shareholder register entirely because of exercise of dominant shareholder force. It is hard to know how to protect people from the effects of the contracts that they have signed, but a start would be making 'legal aid' or other grant assistance available to those who invested under EIS schemes for pursuit of shareholder oppression cases.

A final suggestion, as discussed above, is to reduce the number of schemes that governments set up to support new companies, industries, projects etc., and as a corollary let the schemes run long enough to be a success. I have documented the number of such schemes in the UK: the UK is not alone in this. In 2005 France set up 67 regional partnerships to develop science and technology 'engines' to create jobs [306]. This is a geographic area that is smaller than California in size and economy. Although the overall budget was enormous, each region would receive only $7.5 million/year, shared by many partners in the region. Even if concentrated into one company, that is enough to fund it through just one VC round. There are political reasons why funds are distributed in Europe in this way, and those reasons might be insuperable, but I would urge politicians to at least try to overcome them.

13
What to Do about It: Business

The wheels of government grind exceeding slow: while European states and regions modify the underlying legal and economic structure that encourages the VCs' business model, we can ask what other business people might do to avoid its fallout.

The most clearly affected by the VC business model are institutional investors who invest in the VC funds. We should ask what investors in specialist biotech VC funds might learn from the reality of those funds' activities. There are a number of obvious lessons to be learned from the statistics of what actually happens in VC investing in biotechnology, which I will only skim here as this book does not pretend to be an expert analysis of how pension funds should be managed.

There has been substantial and sustained investment in VC. The National Association of Pension Funds (NAPF), reviewing their members' investment profile [22, 307], noted a major shift towards VC investment in the late 1990s: many pension funds had surpluses, and their focus was on how to invest in new areas such as VC. A variety of factors have been blamed for subsequent scheme deficits, the ones mentioned in the NAPF report being 'increased longevity and adverse market movements' [307] (a curious choice of excuses – life expectancy at age 65 in England and Wales has increased from 23.2 years in 2001 to 23.8 years in 2006 for men, 26.3 to 26.8 for women, hardly enough to bankrupt a pension scheme). 54% of funds by asset class were invested in VC in 2001: it is likely that the poor performance of VC funds is at least a contributory reason to the funds' subsequently poor performance and consequent deficit status. In 2002–4 sentiments moved away from VC as funds focused on 'deficit correction'. Despite this, funds continued to invest in VC: at the end of 2006 many of them still had holdings in VC funds: 19% in 2006, versus 15% in 2005 and 9% 2004.

Investment in VC therefore appears to be here to stay. Better mechanisms and criteria for evaluating VC should therefore be put in place if this continued investment is not to result in continued losses of the type illustrated in Figure 3.2. Among such criteria could be more objective valuation of portfolios. The 'EVCA Guidelines' [68] for valuing investor portfolios are written by private investor interest groups, not by the companies that invest in them, an almost unique state of affairs. Clearly from the *observed* performance of the funds, even under these rules, the claims that biotechnology VCs make money for their investors are hard to sustain.

The principle problem with the VC business model described in Chapter 11 is that the value *to the VC* of raising a fund is greater than the value *to the VC* of effectively managing previously raised funds. VCs will therefore act as intelligent agents and abandon the effective management of old funds, directing their management efforts to developing their prior investee companies as marketing material for their new fund-raising. This can be overcome by reducing VC dependence on management fees and increasing their motivation through profit share. Obviously, there is a trade-off here: with no management fee the VC management group would have to have enough wealth to maintain itself through ten years' operation before it can earn a profit, and if it can do that it is more likely to be an investor in VC funds, not a manager of them. However, it is clear from the data that management fees are too high, and profit motivation too low, to motivate European VCs to behave appropriately. A better balance should be found.

The other set of institutions affected by the current VC model, and the state support that favours it, are the Universities. In Europe Universities are almost entirely funded by the state in various routes, with a minority of the cost of their research activities coming from charities and industry. Their motivation to form spin-out companies is the result of state pressure on the Universities to do so – there is otherwise no rational reason why an institution dedicated to research, teaching and carrying the light of civilisation into the future should be concerned with writing business plans.

There is a continuing debate as to whether Universities, or 'academic' research institutions generally including Universities, hospitals, specialist institutes and government centres, should put significant effort into earning money through commercial activities. The view of a vocal group is that these priorities distort the research agenda in a way that jeopardises the long-term value of research by focusing solely on what can be exploited today [308]. My personal view is that this viewpoint has substantial validity. Apart from issues of moral and social corruption, having 'tick boxes' in Research Council grant applications where applicants can indicate whether research has commercial potential, or has a commercial partner willing to write a letter saying that the research is of potential value, is to forget that almost all truly valuable discoveries in the life sciences came from people who were looking for something else entirely. On a broad, almost philosophical front, therefore it would be more productive for the economy in the long term for those funding academics to take their eyes off the glittering goals of finding patentable inventions and trying to turn them into drugs, and back on finding out how life works. This is how the science that was behind the most successful biotech companies of all time was done – one could, if one wanted to, draw a graph showing the inverse relationship between the growing 'applied science' focus of science ever since and the fall in the average value for funding of biotech companies in Europe.

This is not to say that academics, or their employer institutions, should not patent their inventions when they make them. For reasons mentioned above, without patents to protect them, inventions in the life sciences are unlikely to be developed and applied to the wider world as commercial products. But patenting an invention discovered by visionary, 'blue skies' research is not the same as demanding that that research be targeted at products known to be patentable before the project is started.

More practically, we can ask Universities to give more thought as to whether new start-up companies are the best way to exploit or apply that science in the biotechnology field. It is widely understood that the 'spin-out' company is only one appropriate route for commercialisation, and rarely the best [284–6, 289]. However, start-ups look like less work and more profit (if the inflated claims of the benefits of start-ups are believed), and have been a performance target for technology transfer offices in many European Universities.

Deakins advises that, rather than launch a project into its own spin-out, the commercial potential of the project is evaluated in a pre-commercialisation programme, which can have any legal form that is appropriate but does not seek to ape the management and control structures of a full-scale R&D company, with the burdensome overhead that VCs typically expect be in place [80]. My observation is that several US Universities have mechanisms for doing this in place: European Universities should realise that the effort involved in setting up and financing a new company is actually very substantial, the likely returns very small (see Chapter 3), and as a consequence they would be better advised to pursue a non-corporate project with a view to licensing than pursuing the spin-out route.

In fact, the reason for seeking commercial routes for academic research should not be *financial* returns at all. The real business of Universities is research and education, as we have noted – if they were effective industrial R&D units, then industry would fund them for the economic return they would make. The value of seeking commercial exploitation of research is the enhancement it gives to both research and training of the science base: training people, generating basic understanding and broadening research agendas to include other types of problem. There is a major education gap in the *practicalities* of doing good science in the UK [69], which reflects not just on the specifics of company failure but also on the generalities of research that enabled research to be applied in the future. Simcha Yong,[1] studying the role of Universities in technology transfer, has found that many leading academics in the US who have started companies said that they have gone back to academia with good ideas for their academic research – both topics and organisation – derived from their industrial experience. Europe as a whole is lagging behind on this learning experience [309]. This, and not investment schemes, is the real value of a university participating in academic entrepreneurship. Generating weak spin-outs that can be used to bolster VCs' ability to earn management fees does not meet this class of objectives.

And finally, the entrepreneurs themselves.

The classic work on entrepreneurship by Schumpeter presents a vision of the entrepreneur as hero, breaking the boundaries of convention and reaping the benefits. This plays well to the mythology of biotechnology, and to scientists and businessmen who see the grand, world-altering opportunities in new technology and invest their careers in making them happen. Biographical sketches of successful entrepreneurs play on this image (see, for example, [162, 310–11]).

The reality as described here is quite different. It is well established that the founders of VC-backed companies rarely make money from their creations – indeed in the UK, academic founders are often better off exploiting their discoveries by writing books about them than by starting companies [63–4] (see Figure 3.5). This is so at variance with the mythology of the industry that many entrepreneurs of my personal acquaintance have felt seriously let down by someone in the last decade, and VC is usually cited as the primary cause of the problem. 'Ripped off' is the phrase that most often is used, which may be harsh, and is technically incorrect. Despite some investigations no criminal act has been proven against a European venture management group (although news stories linking 'fraud' and 'biotech' are predominantly about financial fraud rather than scientific fraud, even though there are probably 1000 VCs in the Western world to 100,000 life scientists [312]).

Entrepreneurs with experience have been deterred from seeking VC investment by this, and hence from exploring business opportunities that need VC investment. Over half of the entrepreneur respondents in a study said that the past effects of VC investment, in particular the combination of investment terms such as preference shares and anti-dilution provisions, extremely tight VC control and VC drive towards business sub-optimal business models had discouraged them or close colleagues from starting new entrepreneurial ventures [148]. Even for companies that are successful (in VC terms) and achieve an IPO, VC involvement does not *increase* company value, only decrease the founder and entrepreneur share of it. Florin [184] observes that

> Results suggest that founders motivated primarily by wealth creation and those motivated by remaining in control of their ventures should, in both instances, minimize VC backing when taking their ventures public.

But VCs state that their sole motivation is financial reward – career development for the entrepreneur or social welfare is not on their agenda. So the very entrepreneurs that are most 'investable' are the ones least likely to be happy with VC involvement. The industry-damaging effect of this is that experienced, successful entrepreneurs will be taken out of the pool of company creators, and only the naive or the inexperienced will start companies up [71], increasing the distance between European biotech and that in the US [95]. As we have discussed, this actually benefits VC under the current model, which again illustrates why the current model should be reconsidered.

This applies not only to the entrepreneurial management of start-up companies, but also to Universities and to personal investors in those companies – the 'Business Angels'. As a result of seeing the lack of success of their start-up companies, Universities have pulled back from patenting inventions in the last few years.

Overall patent publications levels are now down to the levels of the mid 1990s, before the boom in VC investment in university spin-outs, reflecting (and presumably caused by) a loss of belief that there is any point in pursuing commercial value from research (Figure 13.1). Studies

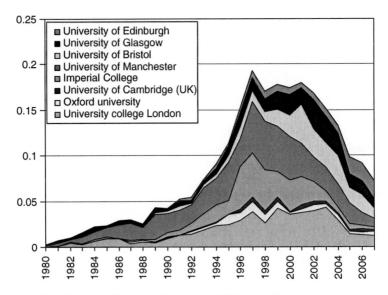

Figure 13.1   Patent publications from some UK Universities
Note: Fraction of all UK patent publications coming from eight selected UK Universities. Note that Oxford and Cambridge have anomalously low figures due to unusual IP ownership and filing organisations.
Source: Data from Espacenet, search February 2008.

such as mine [22] on the low rewards academics and their institutions make from start-up companies merely quantify this trend, which was well under way at the start of the century. It is clearly a bad thing when Universities have decided that there is no point taking out legal protection for their inventions, protection which (as we have noted above) is essential for their application.

Lastly, the 'Business Angels', private investors in start-up companies who are the real drivers behind many innovative start-ups. This book has scarcely mentioned Business Angels because, although often aggregated with early stage VCs as sources of investment, their motivations and business model are completely different. They draw no management fee, invest their own money rather than other people's and are motivated solely by profit on their investment: consequently the counter-productive business model described in Chapter 11 does not apply to them. However, the fallout from that business model does apply to them when they invest in a company that subsequently requires VC investment. The investment structures used by VCs reduce the likely returns to an early stage investor as well as entrepreneurs, and consequently 'Business

Angels' are much less likely to be involved in company creation or seeding in any industry that requires VC investment [148]. An extreme but nevertheless illustrative example was a comment by one VC in 2002 (with apparent seriousness) that if entrepreneurs brought their companies to him for investment, he would only consider taking them off their hands for nothing if they had at least £2 million of the entrepreneur's money invested in them already [313], that is, any new company had a negative inherent value, and needed investment to make it worthless. Only the foolish would invest before the VC under these circumstances. In part, this is a reflection of a wider antipathy between angels and VCs [301], the former seeing investment by the latter (rightly, in this industry) as being incompatible with the angel making a return from investment. The requirement for biotechnology companies to seek further funding means that angels will inevitably suffer the exploitative terms imposed by VC, and as a result suffer loss of their investment even if the company is successful: consequently, they will not invest in biotechnology.

This is clearly damaging, as overall angels support about eight times as many companies as VCs in the UK [9, 110], possibly even more in the US [314] and these are the companies that are the 'feedstock' for the VC's investment business[21], so, in Nelson's words, VCs have 'poisoned their own seed corn' by cutting off the supply of early stage companies in which they can invest [315]. A common complaint from VCs reported by Chemistry Leadership Council (CLC) is that there are too few serial entrepreneurs in Europe, (see [112]), but when fund-raising each round takes nearly a year, rounds are small so you need several, and at the end VC practices perceived as 'predatory' by entrepreneurs and angel investors result in negligible return on investment for the latter [112, 148], then of course those people are not going to come back to do it again. The lack of serial entrepreneurs is an effect of VC policies, not a cause of them. As a consequence of this, the deterrent effect of the fear of failure is actually increasing in UK entrepreneurs as they gain experience of VC-funded companies, not decreasing [48].

The relationship between founder and investor can degenerate under these conditions to become almost pathologically confrontational [153] Clearly open warfare within a company is unlikely to benefit anyone, and some responsibility for such a situation must lie with the entrepreneurial or management team. But a necessary, and often sufficient, precondition for it is the belief, amply supported by empirical evidence that we have examined, that once a VC comes into a company any hope that the entrepreneur has of adding value to the company or getting value from it is gone.

Carl Feldbaum, then CEO of the BIO, the US industry organisation, was sufficiently concerned at the fate of entrepreneurs to draft a ten-point 'bill of rights for entrepreneurs' in 1998 [316], which, as well as including 'rights' concerning IP law, state funding for basic research and other things with which Europe would have no problems, said that entrepreneurs had a 'right' such that

> governments should ensure that securities regulatory agencies adopt financial market regulations that reduce the burden for entrepreneurs and entrepreneurial firms in private placements, initial public offerings, and secondary offerings of stock on the public markets.

This obviously has not happened in the following decade in Europe.

What can European biotechnology entrepreneurs do about this? The answer to this question used to be that there were just three solutions: put up with it (i.e. draw a salary and abandon hopes of wealth or influence), go to the US, or give up. Luckily there is an emerging fourth answer – seeking European financing for high-risk, high-growth companies outside the VC funding stream. We have touched on the rise of the AIM in London as a source of finance for very early stage companies. This is portrayed as an alternative exit route for VCs who are stymied in their desire to get high value exits for low value companies on the main London Stock Exchange (LSE). But in reality it has become a funding route for companies that is parallel to VC: companies such as Alizyme, VasTox, Synairgen and others have floated on AIM without any prior VC funding, allowing their founders and initial investors to retain significant fractions of the equity in the companies, and to grow the companies following a valid long-term business model rather than one aimed at producing backing material for VC fund-raising.

Hedge funds are starting to move into the early stage investment space in biotechnology as well. Untrammelled by the VC business model, they can invest in ordinary stock and seek to gain rapid, high-value exits as a small part of a larger growth portfolio. Hedge funds are usually too large to bother with investments below $10 million, but some specialists have seen this as a niche, and have been welcomed enthusiastically by entrepreneurs, some dumping VC funding at the last minute in preference to Hedge fund financing. Whether Hedge funds will prove any less destructive in the long term remains to be discovered.

By far the most attractive route, though, is to reengineer the start-up's business model to avoid the need for VC in the first place. How this is done depends on the specifics of the entrepreneur and the technology: this, however, is the essence of entrepreneurship – finding ways of bridging impossible chasms to achieve commercial success. The critical question the entrepreneur, whether scientist or salesman, needs to ask is – 'If I go down a VC route, what will happen?' This book provides a lot of data to help you answer that. If what is likely to happen means that VC funding is just not economically worth pursuing, then you *have* to either find another business model or give up. The true entrepreneur will find another way.

Notes

1 Genes and Money in Europe: Why Did It Fail?

1. I know someone who did this.
2. Bonsai trees are miniature trees cultivated from full-sized plants by taking a seedling and systematically restricting its growth by clipping its shoots and roots and growing it in a tiny, confining pot. Bonsai trees are wonderful creations, but you cannot build houses or ships from the timber.
3. In mitigation, I also offer the evidence of my complete collection of *Buffy the Vampire Slayer* DVDs.
4. European Venture Capital Association.
5. Bio Industry Association.

2 The Biotechnology Industry and Venture Capital

1. For relatively accessible technical introductions to the science behind biotechnology, I can recommend Bains [3], Smith, J.E., [4], Ratledge, C. and B. Kristiansen, eds [5]. For a light introduction to the genome project that featured so prominently in biotechnology between 1998 and 2002, see Ridley, M., [6].
2. As evidenced by their respective company websites, accessed in 2007.

3 Europe Falls Behind

1. I amazed my colleagues by using an 'electronic mail' system to apply for US postdoctoral research positions in 1982.
2. BVCA – British Venture Capital Association. EVCA – European Venture Capital Association. AFIC – Association Française des Investisseurs en Capital.
3. The observant reader will note that I am writing a book about this.
4. This is a small sample of three books and three companies, but I believe is representative.

4 Why Does Europe Do So Poorly? Some Inadequate Explanations

1. Coombs and Alston [79] list nearly 600, but a number of these are large manufacturing concerns, including major pharmaceutical companies, wholly owned subsidiaries of overseas companies, or companies producing non-industry-specific equipment (such as valves or pumps) that happened to be sold to biotech companies. These companies are excluded from analysis here.
2. The salary of the staff carrying out this deal flow analysis is, in my informal and anecdotal experience, between 5% and 15% that of the VC partners.

3. The 'Biopolis' in Singapore alone has facilities for UK companies Cyclacel, Paradigm Therapeutics, Stem Cell Sciences and attracted David Lane (founder Cyclacel), Alan Coleman (formerly with PPL Therapeutics), Alex Matter (formerly with Novartis) and Axel Ullrich (founder of several companies) as senior European appointees (http://www.biomed-singapore.com).
4. My own start-up, Amedis Pharmaceuticals, is typical. Based in Cambridge, UK, at start-up it had its principal research base at Imperial College in London, and drew its IP from Germany and New York.

5   Underinvestment in European Biotechnology

1. A drug that boosts the body's production of blood, and Amgen's first multibillion-dollar blockbuster product.

6   Public Markets and Underinvestment

1. Unpublished research done in collaboration with Cambridge University.

8   VC Effects on Business Efficacy

1. This is not the place I point out how clever I am – I have founded two drug discovery companies, sit on the boards of others and am a dedicated fan of this area of science. But from an objective economic point of view, it is not a terribly bright thing to do.
2. www.abcam.com.

9   Investor Blockade of Business

1. Note that a £20 million IPO does not mean that this is the maximum the VCs could have made. If the stock rose after IPO, their exit value would have risen with it, and as the company would have raised more money at IPO to drive product development forward, this was quite a likely outcome.
2. Arakis Ltd. Extraordinary General Meeting resolution and documents lodged at Companies House, UK (www.companieshouse.co.uk), document number 3911256.

10   Pomp and Circumstance

1. Luton is a notoriously 'un-prestigious' address for the Cambridge biosciences cluster's inhabitants.
2. Several commentators have stated that 'VCs no longer want quantity in the IP section – they look for quality.' What this translates to in practice, though, is that investors require both quantity and quality, which is even more expensive to achieve.
3. A recent detailed analysis of patents in US biotech companies by Parida et al (Nature Biotechnology 2008, 26: 763–766) Shows that, while patents are

essential for some biotech businesses, there is essentially no correlation between the *number* of patents and the success of the business in developing and launching products.

11   The Real VC Business Model

1. Rounds on finance within 12 months, and usually within six months, of an IPO, made to make companies more 'solid' for the stock market rather than for business or product development.
2. Just in case the reader thinks that the author is bashing everyone except himself, I should state here that I spent eight years as a consultant before joining a VC company. I am not happy with the conclusion, summarised here, but these are the facts, and in the spirit of the book I am letting the facts speak.
3. Expressed Sequence Tag.
4. He is alleged to have said that if the weather forecasts for air raids were 45% accurate, if he did exactly the opposite of what the weather forecasters suggested he would be right 55% of the time.
5. The biochemically informed reader will see a parallel with competitive binding here, and how it is easier to find a competitive antagonist to a very 'promiscuous' site than to a discriminating one.

12   What to Do about It: Government

1. Including at least one of mine, in case the reader thinks this section is 'sour grapes'.
2. Eloquently expressed in the UK political comedy series *Yes, Prime Minister* ([303] Lynn, J. and A. Jay, *Yes, Prime Minister*), this is 'We must do something. This action is something. Therefore we must do this'. The majority of twentieth century political decisions can be ascribed to this logic, and the current century shows no sign of being better.
3. In the event (2008) this new initiative, like so many such announcements, resulted in no actual changes or investment.

13   What to Do about It: Business

1. Cambridge University PhD student, in conversation with the author.

References

1. Metcalfe, J.S., *Restless capitalism, experimental economics*, in *New technology-based ventures in the new millennium*, W. During, R. Oakey, and M. Kipling, eds 2001, Pergamon: Amsterdam. pp. 4–16.
2. Graham, P., *A unified theory of VC suckage* http://www.paulgraham.com/venture capital.html (produced by Paul Graham, accessed 24 March 2005).
3. Bains, W., *Biotechnology from A to Z*. 3rd ed. 2004, Oxford, UK: Oxford University Press. 0–19-852498-6.
4. Smith, J.E., *Biotechnology*. 3rd ed. 1996, Cambridge, UK: Cambridge University Press.
5. Ratledge, C. and B. Kristiansen, eds *Basic Biotechnology*. 2nd ed. 2001, Cambridge University Press: Cambridge, UK.
6. Ridley, M., *Genome. The autobiography of a species*. 1999, London, UK: Fourth Estate.
7. Ernst and Young, *Beyond borders: A global perspective – Biotechnology industry report 2006*. 2006, Ernst and Young Ltd. (New York, May 2006).
8. Bains, W., *Open Source and Biotech*. Nature Biotechnology, 2005. 23(9): p. 1046.
9. Mason, C. and R. Harrison, *The size of the informal venture capital market in the United Kingdom*. Small Business Economics, 2000. 15: pp. 137–148.
10. Bains, W., *Failure rates in drug discovery and development: Are we getting any better?* Drug Discovery World, 2004. 2004 (Fall): pp. 9–18.
11. Smith, J.G. and V. Fleck, *Strategies of new biotechnology firms*. Long Range Planning, 1988. 21(3): pp. 51–8.
12. Gompers, P.A. and J. Lerner, *The Venture Capital Cycle: 2nd edition*. 2004, Cambridge, MA: MIT Press.
13. Sætre, A.S., *Entrepreneurial perspectives on informal venture capital*. Venture Capital, 2003. 5(1): pp. 71–94.
14. Amis, D. and H. Stevenson, *Winning Angels: The seven fundamentals of early stage investing*. 2001, London, UK: Prentice Hall.
15. Bradford, T.C., *Evolving symbiosis – venture capital and biotechnology*. Nature Biotechnology, 2003. 21(9): pp. 983–4.
16. Dibner, M.D., M. Trull, and M. Howell, *US Venture capital for biotechnology*. Nature Biotechnology, 2003. 21(6): pp. 613–17.
17. Lee, D.P. and M.D. Dibner, *The rise of venture capital and biotechnology in the US and Europe*. Nature Biotechnology, 2005. 23(6): pp. 672–6.
18. Kenny, M., *Biotechnology: The university: Industrial complex*. 1986, New Haven: Yale University Press.
19. Martin, P. and S. Thomas, *The 'commercialization gap' in gene therapy: Lessons for European competitiveness*, in *Biotechnology and Competitive Advantage*, J. Senker and R. van Vliet, eds 1998, Edward Elgar: Cheltenham, UK. pp. 130–55.
20. Manigart, S., M. Wright, K. Robbie, P. Desbrières, and K. De Waele, *Venture capitalists' appraisal of investment projects: An empirical European study*. Entrepreneurship Theory and Practice, 1998. 21(4): pp. 29–44.

21. Murray, G.C., *Evolution and change: An analysis of the first decade of the UK venture capital industry*, in *Venture Capital*, M. Wright and K. Robbie, eds 1997, Dartmouth: Aldershot, UK. pp. 54–82.
22. Bains, W., *How academics can make (extra) money from their science*. Journal of Commercial Biotechnology, 2005. 11(4): pp. 1–11.
23. European Investment Fund, *Annual Report 2004*. 2004, European Investment Fund (Luxembourg).
24. Saviotti, P.P., P. -B. Joly, J. Esrades, S. Ramani, and M. -A. de Looze, *The creation of European dedicated biotechnology forms*, in *Biotechnology and Competitive Advantage*, J. Senker and R. van Vliet, eds 1998, Edward Elgar: Cheltenham, UK. pp. 68–88.
25. Tegenero, *Tegenero Immuno Therapeutics (website)* http://www.tegenero.com/about_us/index.php (produced by Tegenero, accessed 21/02/2006).
26. Ernst and Young, *Biotech 91: A changing environment*. 1990, Ernst and Young (SanFrancisco, 1 February. 1990).
27. Ernst and Young, *Biotech 97: Alignment*. 1996, Ernst and Young (San Francisco, CA, 1 April 1996).
28. Ernst and Young, *Integration: Ernst and Young's Eighth annual European life science report 2001*. 2002, Ernst and Young (London, 2002).
29. Bioscan database, 1995, AR, USA: Oryx Press.
30. Bioscan database, 1998, AR, USA: Oryx Press.
31. Walsh, G., *Biopharmaceutical benchmarks 2006*. Nature Biotechnology, 2006. 24(7): pp. 769–76.
32. Lähteenmäki, R. and S. Lawrence, *Public Biotechnology 2005 – the numbers*. Nature Biotechnology, 2006. 24(6): pp. 625–41.
33. Lawrence, S., *Biotech as an employer*. Nature Biotechnology, 2007. **25**(1): p. 12.
34. Chipman, A., *Fatal Attraction*. Nature, 2005. 436(7051): pp. 624–5.
35. Johnstone, B., *Britain's stake in biotechnology*, London Times, (London), 12/11/1980, p 26.
36. Anon, *Grand Met puts $10M into genetic engineering*, Times of London, (London), 25/3/1980, p. 19.
37. Cookson, C., *The investors' honeymoon with biotechnology comes to an end*, Times, (London), 25 October 1982, p. 16.
38. Tran, M., *Market Marshalls its forces in the battle against AIDS*, Guardian, (London), 13 April 1992, p. 11.
39. Global Business Insights, *The top ten biotechnology companies*. 2005, Global Business Insights, (London, UK).
40. Anon, *Synthetic blood clot agent to be produced*, Times of London, (London), 24/11/1982, p. 2.
41. Anon, *Genetic engineers prepare for battle*, Times of London, (London), 21/3/1983, p. 16.
42. Thakor, A.V., *Comment on Trester*. Journal of Banking and Finance, 1998. 22(6–8): pp. 700–1.
43. Allen, M., *Where even the small firm can find finance*, London Times, (London), 9/5/1972, p. 6.
44. EVCA, *European Venture Capital Association web site* http://www.evca.com/html/home. asp (produced by European Venture Capital Association, accessed 16 February 2007).

45. Wright, P., *Genetic engineering's industrial heartland, London Times*, (London), 7/8/1981, p. 19.
46. Fairtlough, G., *Creative Compartments: A design for future organization*. 1994, London, UK: Adamantine Press.
47. Walsh, V., J. Niosi, and P. Mustar, *Small firm foundation in biotechnology: A comparison of Britain, France and Canada*. Technovation, 1995. 15(5): pp. 303–27.
48. Harding, R., *Global Entrepreneurship Survey 2004*. 2005, London School of Economics (London).
49. Fishlock, D., *Why industry is taking more notice, London Times*, (London), 27/12/1984, p. 9.
50. Fishlock, D., *The reluctant entrepreneurs, London Times*, (London), 4/10/1982, p. 17.
51. Wright, P., *Biotechnology: 'Genetic engineering' creates a new industrial revolution, Times*, (London), 26 February 1981, p. 4.
52. Mason, C., S. Cooper, and R. Harrison, *The role of venture capital in the development of technology clusters: The case of Ottawa*, in *New technology-based firms in the new millennium*, vol II, R. Oakey, W. During, and S. Kauser, eds 2002, Pergamon: Amsterdam. pp. 261–78.
53. Avnimelech, G. and M. Teubal, *Venture capital start-up co-evolution and the emergence and development of Israel's high-tech cluster*. Econ. Innov. New. Techn., 2004. **13**(1): pp. 33–60.
54. Anon, *Cleantech firms attracted $1.3bn of global VC investment in 2006*, in *Freeradicals*. 2007. p. 69.
55. Lynn, M., *Drug firms seek cash injection abroad, Sunday Times*, (London), 7 March 1993, p. 15.
56. Woolcock, K., *Bio firms looking abroad for funds, Mail on Sunday*, (London), 15 March 1992, p. 74.
57. Arnold, M., *IDM's flight a 'wake-up call' for Europe, Financial Times*, (London), 21 October 2005, p. 20.
58. Pollack, A., *U.S. Finance pulls biotech across seas, New York Times*, (New York), 12 July 2006, p. 12.
59. Pollack, A., *Europe's Biotech 'immigrants' to America, New York Times*, (New York), 11 July 2006, p. 11.
60. Freel, M.S. and P.J.A. Robson, *Small firm innovation, growth and performance*. International Small Business Journal, 2004. 22(6): pp. 561–75.
61. Harris, D., *Fast cash flow that helps industry grow, Times*, (London), 10 May 1985, p. 15.
62. Gompers, P.A., *Venture capital growing pains: Should the market diet?* Journal of Banking and Finance, 1998. 22(6–8): pp. 1089–104.
63. Bains, W., *Recognising the value of invention*, in *IP Review*. 2006. p. 6.
64. Atchison, K., *Reply to Bains*, in *IP Review*. 2006. p. 6.
65. Bygrave, W., N. Fast, R. Khoylian, L. Vincent, and W. Yue, *Early rates of return of 131 venture capital funds started 1978–84*. Journal of Business Venturing, 1989. 4(2): pp. 93–105.
66. Manigart, S., P. Joos, and D. DeVos, *Performance of publically traded European VC companies*, in *Frontiers of Entrepreneurial Research*, W. Bygrave, et al., eds 1992, Wellesley: Boston, MA.

67. Tyebjee, T.T. and A.V. Bruno, *A model of venture capitalist investment activity*, in *Venture Capital*, M. Wright and K. Robbie, eds 1997, Dartmouth: Aldershot, UK. pp. 103–18.

68. AIFC, EVCA, and BVCA, *International provate equity and venture capital valuation guidelines*. 2006, International Private Equity and Venture Capital valuation Board (Brussels, 7 January 2007).

69. Bains, W., *Measurements for Biotechnology: Advice and training*. 2007, Measurements for Biotechnology Initiative (LGC/Department of Trade and Industry) (London).

70. Booth, B., *Early-stage returns?* Nature Biotechnology, 2006. 24(11): pp. 1337–2340.

71. Sahlman, W.A., *The structure and governance of venture-capital organizations*, in *Venture Capital*, M. Wright and K. Robbie, eds 1997, Dartmouth: Aldershot, UK. pp. 3–51.

72. 3i Group, *3i Group plc annual report and accounts 2006*. 2006, 3i plc (London, UK).

73. GIMV, *GIMV annual report 2005*. 2006, GIMV (Den Haag, The Netherlands, 2006).

74. Shepherd, J., *Wealthy sultans of spin-off cash in on their ideas*, Times Higher Educational Supplement, (London), 20 January 2006, pp. 8–9.

75. Count, M., *Commercial Breaks for Professors who are not so nutty*, London Times, (London), 24/11/2001, pp. 52–3.

76. Edwards, M.G., *Academic equity in start-up companies: A retrospective look at Biotech IPOs* http://www.recap.com/consulting.nsf/0/7AD3CD2D5E2D3C1788 256FA4005D6596/$FILE/AUTM0205.pdf (produced by Recombinant Capital, accessed 22/4/2005).

77. Gregory, W.D. and T.P. Sheahen, *Technology transfer by spin-off companies vs licensing*, in *University spin-off companies: Economic development, faculty entrepreneurs and technology transfer*, A.M. Brett, D.V. Gibson, and R.W. Smilor, eds 1991, Rowman and Littlefield Publishers: Savage, MD, USA. pp. 133–52.

78. Quested, T., *Evans steps up Merlin fundraising despite SFO probe* www.businessweekly.co.uk (produced by Business Weekly, accessed 12 October 2005).

79. Coombs, J. and Y.R. Alston, *The International Biotechnology Directory* 1988, Basingstoke: Macmillan Publishers.0–333–43726–8.

80. Deakins, D., *Entrepreneurship and Small Firms*. 2nd ed. 1999, London: McGraw Hill.

81. DeYoung, G., *University and Industry in agreement*. Bio/Technology, 1988. 6(8): pp. 906–10.

82. Bunting, S., *Life's rich opportunity*, Financial Times, (London), 4th July 2002, p. 12.

83. Klausner, A., *DNAX keeps a scholar's focus*. Bio/Technology, 1988. 6(4): pp. 373–8.

84. Berry, M., *UK Managers worst in Europe for staff communication*, in *Personnel Today* 2005. pp. 11–12.

85. Trompenaars, F., *Riding the Waves of Culture*. 1997, London, UK: Nicholas Brealey.1857881761.

86. Hendry, C. and J. Brown, *Local skills and knowledge as critical contributions to the growth of industry clusters in biotech*, in *New technology-based firms in the new millennium*, W. During, R. Oakey, and M. Kipling, eds 2000, Permagon: Oxford. pp. 127–40.

87. Klausner, A., *Many angles to competitive analysis.* Bio/Technology, 1987. 5(11): p. 1129.

88. Thomas, E., *Hamming's 'open doors' and group creativity as keys to scientific excellence: The example of Cambridge.* Medical Hypotheses, 2008. 70(3): pp. 473–7.

89. Riquelmo, H. and J. Watson, *Do venture capitalists' implicit theories on new business success/failure have empirical validity?* International Small Business Journal, 2002. 20(4): pp. 395–420.

90. Dimov, D.P. and D.A. Shepherd, *Human capital theory and venture capital firms: Exploring 'home runs' and 'strike outs'.* Journal of Business Venturing, 2005. 20(1): pp. 1–21.

91. Powers, J.B. and P.P. McDougall, *University start-up formation and technology licensing with firms that go public: A resource-based view of academic entrepreneurship.* Journal of Business Venturing, 2005. 20: pp. 291–311.

92. Mitchell, P., *Europe caught in innovation quagmire.* Nature Biotechnology, 2005. 23(9): p. 1029.

93. Critical I, *The equity gap: European and US biotechnology 2001–2.* 2004, Bioindustry Association (London).

94. Critical I, *Comparative Statistics for the UK, European and US Biotechnololgy sectors: Analysis year 2003.* 2005, Department of Trade and Industry (London, February 2005).

95. Bains, W., *What you give is what you get: Investment in European biotechnology.* Journal of Commercial Biotechnology, 2006. 12(4): pp. 274–83.

96. Hogdson, J., *Germany hit badly by investor disinterest.* Nature Biotechnology, 2002. 20(6): p. 531.

97. Lawrence, S., *Tech transfer – slow progress.* Nature Biotechnology, 2005. 23(1): p. 12.

98. Capart, G., *Innovation from public research in Europe,* in *From IP to IPO: Key issues in commercialising university technology (Supplement to Intellectual Property Management magazine),* J. Wild, ed. 2005, Intellectual Property Management magazine: London. pp. 8–10.

99. Newmark, P., *Low pay, low profits in United Kingdom.* Bio/Technology, 1988. 6(3): p. 249.

100. McCormick, D., *1988 compensation survey results.* Bio/Technology, 1988. **6**(9): pp. 1013–18.

101. Anon, *Biotech hamstrung by investment strategies.* Chemistry and Industry, 2004. 2004(1): p. 5.

102. Mitchell, P., *Next-generation monoclonals less profitable than trailblazers?* Nature Biotechnology, 2005. 23(8): p. 906.

103. Wiles, M. and P. Andreassen, *Monoclonals: The billion dollar molecules of the future.* Drug Discovery World, 2006. 2006 (Fall): pp. 17–23.

104. Kola, I. and J. Landis, *Can the pharmaceutical industry reduce attrition rates?* Nature Reviews Drug Discovery, 2004. 3(8): pp. 711–15.

105. Klausner, A., *Genentech's busy lawyers.* Bio/Technology, 1988. 6(5): p. 472.

106. Klausner, A., *Varying views of Genentech's litigation.* Bio/Technology, 1988. 6(6): pp. 646–7.

107. Bains, W., *Litigation escalates for patents on a chip.* Nature Biotechnology, 1997. 15(5): p. 406.

108. Gompers, P.A., *Optimal investment, monitoring and the staging of venture capital*, in *Venture Capital*, M. Wright and K. Robbie, eds 1997, Dartmouth: Aldershot, UK. pp. 285–313.
109. Sweeting, R.C., *UK Venture capital funds and the funding of new technology-based businesses: Process and relationships*, in *Venture Capital*, M. Wright and K. Robbie, eds 1997, Dartmouth: Aldershot, UK. pp. 316–47.
110. Reichhardt, T., *Touched by an angel*. Nature, 2005. 435(7043): pp. 734–5.
111. Sheridan, C., *German biotech gets a second chance*. Nature Biotechnology, 2004. 12(12): pp. 1414–15.
112. Chemistry Leadership Council, *Chemical science spin-outs from UK Universities: Review of critical success factors*. 2005, Royal Society of Chemistry and the Chemistry Leadership Council (London, UK, 2005).
113. Pearson, S., *European Biotechs face a funding crisis*. Genetic engineering news, 2006. 26(16): pp. 1–10.
114. Mitchell, P., *Could bank loans solve Europe's biotech financing slump*. Nature Biotechnology, 2005. 23(12): pp. 1459–60.
115. Murray, G., *The second 'Equity Gap': Exit problems for seed and early stage venture capitalists and their investee companies*. International Small Business Journal, 1993. 12(4): pp. 59–75.
116. Burnand, A., *The Future is Bright*, 3 April 2006. pp. 17–19.
117. Anon, *The YIC factor*. Nature Biotechnology, 2005. 23(10): p. 1187.
118. Library House, *The false funding gap*. European Venture Capital Journal, 2006. 136: p. 9.
119. Harding, R. and M. Cowling, *Assessing the scale of the equity gap*. Journal of small business and enterprise development, 2006. 13(1): pp. 115–31.
120. Gray, C., *Managing entrepreneurial growth: A question of control?*, in *Small firms: Entrepreneurship in the nineties*, D. Deakins, P. Jennings, and C. Mason, eds 1997, Paul Chapman Publishing: London. pp. 18–28.
121. Gorman, M. and W.A. Sahlman, *What do venture capitalists do?* Journal of Business Venturing, 1989. 4: pp. 231–48.
122. Lawrence, S., *US trails Europe in IPOs*. Nature Biotechnology, 2005. 23(10): p. 1195.
123. Moran, N., *Technology commercialization firms float in UK – but not elsewhere*. Nature Biotechnology, 2007. 25(7): pp. 697–8.
124. Papadopoulos, S., *Evolving paradigms in biotech IPO valuations*. Nature Biotechnology, 2001. 19(Supplement): pp. BE18–BE19.
125. Burrill, G.S., *The beginning of the biotech century?* Modern Drug Discovery, 2000. 3(5): pp. 75–6.
126. Lawrence, S., *Upward Trend in financing continues, but fewer feel flush*. Nature Biotechnology, 2007. 25(1): p. 3.
127. McGully, M.G. *Current Trends in Deals and Financing*. in *GTCbio's Metabolic Diseases World Summit: Partnering and Deal-making summit*. 2005. Burlingame, San Francisco, CA, US.
128. Edelson, S., *Bye-bye boom and bust*. Biocentury, 2006. 14(1): pp. A1–A6.
129. Bains, W., *The long-term value of genomics companies*. Journal of Chemical technology and biotechnology, 2000. 75: pp. 883–900.
130. Aldridge, S., *Business Savvy for UK Biotechs*. Genetic engineering news, 2007. 27(2): pp. 12–13.

131. Sheridan, C., *Germany biotech gets a second change.* Nature Biotechnology, 2003. 21(12): pp. 1414–15.
132. Anon, *Merlin chief warns over funding, Daily Telegraph,* (London), 13 November 2003, p. 35.
133. Dalal, Y. and S. Orn, *Examining an innovative financing alternative for mid-stage biotechs.* 2006, Kellogg School of Management (Evanston, IL, Winter 2006).
134. Dyer, G., *Biotechnology industry facing funding crisis, Financial Times,* (London), 17 November 2003, p. 5.
135. Voison, E., *Biotech's comeback in France.* Bioprocess international, 2007. Supplement: European Conference: pp. 8–13.
136. Pilling, D., *Survey: Life sciences, Financial Times,* (London), 28 October 1999, p. 1.
137. Anon, *All together now?* Nature Biotechnology, 2006. 24(12): p. 1454.
138. Mason, C. and M. Stark, *What do investors look for in a business plan? A comparison of the investment criteria of bankers, venture capital and business angels.* International Small Business Journal, 2004. 22(3): pp. 227–48.
139. Mason, C. and A. Rogers, *The business angel's investment decision: An exploratory analysis,* in *Small firms: Entrepreneurship in the nineties,* D. Deakins, P. Jennings, and C. Mason, eds 1997, Paul Chapman Publishing: London. pp. 29–46.
140. Sahlman, W.A., *How to write a great business plan.* Harvard Business Review, 1997. 1997 (July–August).
141. Föller, A., *Leadership management needs in evolving biotech companies.* Nature Biotechnology, 2004. 20 (Bioentrepreneur supplement): pp. BE65–BE66.
142. MacMillan, I.C., L. Zemann, and P.N. Subbanarasimha, *Criteria distinguishing successful from unsuccessful ventures in the venture screening process,* in *Venture Capital,* M. Wright and K. Robbie, eds 1997, Dartmouth: Aldershot, UK. pp. 123–33.
143. Cardullo, M.W., *Technological entrepreneurism.* 1999, Baldock, UK: Research Studies Press.
144. Oakey, R.P., *Technical entrepreneurship in high technology small firms: Some observations on the implications for management.* Technovation, 2003. 23: pp. 679–88.
145. Walton, A., *Some thoughts on bioentrepreneurship.* Nature Biotechnology, 1999. 16(Supplement): pp. 7–8.
146. Fletcher, L., *Merlin's Maestro.* Nature Biotechnology, 2002. 20(Supplement): pp. BE13–BE14.
147. Chandler, G.N. and S.H. Hanks, *An examination of the substitutability of founders human capital and financial capital in emerging business ventures.* Journal of Business Venturing, 1998. 13: pp. 353–69.
148. Maynard, O. and W. Bains, *Share structure and entrepreneurship in UK biotechnology companies: An empirical study. Journal of Commercial Law Studies,* 2008. 81(1): pp. 1–37.
149. Fried, J. and M. Ganor, *Common stock vulnerability in venture-backed startups.* New York University Law Review, 2006 (May 2006).
150. Wyatt, M., *A study of the dynamic nature of principle agent relationships within the venture capital industry – based on an analysis of venture capital investing in the UK biotechnology industry in 1999 & 2002,* in *Executive MBA.* 2004, University of Warwick: Coventry.

151. Kanniainen, V. and C. Keuschnigg, *Start-up investment with scarce venture capital support.* Journal of Banking and Finance, 2004. 28(8): pp. 1935–59.
152. Cyr, L.A., D.E. Johnson, and T.M. Welbourne, *Human resources in initial public offering firms: Do venture capitalists make a difference?* Entrepreneurship Theory and Practice, 2000. 25(1): pp. 77–91.
153. Loizos, C., *The toxic entrepreneurs.* Venture Capital Journal, 2005. 2005(April 7): pp. 1–3.
154. Rothwell, R., *VC, small firms and public policy in the UK.* Research Policy, 1985. 14: pp. 253–65.
155. Manigart, S., K. Baeyens, and W. Van Hyfte, *The survival of venture capital backed companies.* Venture Capital, 2002. 4(2): pp. 103–24.
156. Busenitz, L.W., J.O. Fiet, and D.D. Moesel, *Reconsidering the venture capitalists' 'value added' proposition: An interorganizational learning perspective.* Journal of Business Venturing, 2004. 19: pp. 787–807.
157. MacMillan, I.C., D.M. Kulow, and R. Khoylian, *Venture capitalists' involvement in their investments: Extent and performance,* in *Venture Capital,* M. Wright and K. Robbie, eds 1997, Dartmouth: Aldershot, UK. pp. 246–65.
158. Aldrich, H.E., A.B. Elam, and P.R. Reese, *Strong ties, weak ties and strangers,* in *Entrepreneurship in a global context,* S. Birley and I.C. MacMillan, eds 1997, Routledge: London. pp. 1–25.
159. O'Regan, N., M. Sims, and A. Ghobadian, *The impact of management techniques on performances in technology-based firms.* Technovation, 2004. 24: pp. 265–73.
160. Barry, C.B., *New directions in research on venture capital finance.* Financial Management, 1994. 23(3): pp. 3–15.
161. Zider, B., *How venture capital works.* Harvard Business Review, 1998. 1998 (December).
162. Fletcher, L., *Sir Chris Evans, the UK's leading bioentrepreneur, reflects on his evolution from lab rat to financial wizard.* Nature Biotechnology, 2002. 20(6): pp. BE13–BE14.
163. Korn/Ferry International, *25th Annual Board of Directors Study.* 1998, Los Angeles, CA: Korn Ferry International.
164. Ferris, S.P., M. Jagannathan, and A.C. Pritchard, *Too busy to mind the business? Monitoring by directors with multiple board appointments.* Journal of Finance, 2003. 58(3): pp. 1087–1111.
165. Ruhnka, J.C., H.D. Feldman, and T.J. Dean, *The 'living dead' phenomenon in venture capital investments,* in *Venture Capital,* M. Wright and K. Robbie, eds 1997, Dartmouth: Aldershot, UK. pp. 350–67.
166. Parjankangas, A. and H. Landström, *How venture capitalists respond to unmet expectations: The role of social environment.* Journal of Business Venturing, 2006. 21: pp. 773–801.
167. Murray, G. and R. Marriot, *Modelling the economic viability of an early-stage, technology focussed, venture capital fund,* in *New technology-based firms in the 1990s, vol 5,* R. Oakey and W. During, eds 1998, Paul Chapman Publishing: London. pp. 97–121.
168. Hellmann, T. and M. Puri, *Venture capital and the professionalization of start-up firms: Empirical evidence.* Journal of Finance, 2002. 57(1): pp. 169–97.
169. Boeker, W. and R. Karichalil, *Entrepreneurial transitions: Factors influencing founder departure.* Acad. Manage. J., 2002. 45(4): pp. 818–26.

170. Bains, W., *When should you fire the founder.* Journal of Commercial Biotechnology, 2007. 13(3): pp. 139–49.
171. Hambrick, D.C. and L.M. Crozier, *Stumblers and start in the management of growth.* Journal of Business Venturing, 1985. 1: pp. 31–45.
172. McCarthy, A.M. and D.A. Krueger, *Changes in the time allocation patterns of entrepreneurs.* Entrepreneurship Theory and Practice, 1990. 15: pp. 7–18.
173. Drucker, P., *Innovations and Entrepreneurship.* 1985, New York: Harper and Row.
174. Clifford, D.K. and R.E. Cavenaugh, *The Winning Performance.* 1985, New York: Bantam Books.
175. Carroll, G.R., *Dynamics of publisher succession in newspaper organizations.* Adm. Sci. Q., 1984. 29: pp. 93–113.
176. Haveman, H.A., *Ghosts of managers past: Managerial succession and organizational mortality.* Acad. Manage. J., 1993. 36: pp. 864–81.
177. Barth, E., T. Gulbrandsen, and P. Schøne, *Family ownership and productivity: The role of owner-management.* Journal of Corporate Finance, 2005. 11: pp. 107–27.
178. Smith, B.F. and B. Amoako-Adu, *Management succession and financial performance of family controlled firms.* Journal of Corporate Finance, 1999. 5: pp. 341–68.
179. McConaughty, D.L., M.C. Walker, G.V. Henderson, and C.S. Mishra, *Founding family controlled firms: Efficiency and value.* Review of Financial Economics, 1998. **7**: pp. 1–19.
180. Barnes, L.B. and S.A. Hershon, *Transferring power in the family business.* Family Business Review, 1989. 2: pp. 187–202.
181. Granlund, M. and J. Taipaleenmäki, *Management control and controllership in new economy firms – a life cycle perspective.* Management accounting research, 2005. 16: pp. 21–57.
182. Willard, G.E. and D.A. Krueger, *In order to grow, must the founder go: A comparison of performance between founder and non-founder managed high-growth manufacturing firms.* Journal of Business Venturing, 1992. 7(3): pp. 181–94.
183. Higashide, H. and S. Birley, *The consequences of conflict between the venture capitalist and the entrepreneurial team in the United Kingdom from the perspective of the venture capitalist.* Journal of Business Venturing, 2002. 17: pp. 59–81.
184. Florin, J., *Is venture capital worth it? Effects on firm performance and founder returns.* Journal of Business Venturing, 2005. 20: pp. 113–35.
185. Parker, S.K. and M. Skitmore, *Project management turnover: Causes and effects on project performance.* International Journal of Project Managament, 2005. 23: pp. 205–14.
186. Edgington, S.M., *What went wrong with Centoxin.* Bio/Technology, 1992. 10: p. 617.
187. Spalding, B.J., *FDA setback flattens Centocor.* Bio/Technology, 1992. 10: p. 616.
188. Scarlett, J.A., *Biotechnology's emerging opportunities: Lessons from the Bauhaus.* Nature Biotechnology, 1999. 17(Supplement): pp. BE13–BE15.
189. Formela, J.-F., *Business models for the biotech entrepreneur.* Nature Biotechnology, 1999. 16(Supplement): p. 16.
190. Stoiber, W., *Consolidation challenges in Europe's life science sector – making it happen.* Journal of Commercial Biotechnology, 2003. 10(1): pp. 29–34.

191. Hodgson, J., *Crystal gazing the new technologies.* Nature Biotechnology, 2000. 18(1): pp. 29–31.

192. Drews, J., *Biotechnology's metamorphosis into a drug discovery industry.* Nature Biotechnology, 1999. 16(Supplement): pp. 22–4.

193. Papadopoulos, S., *Business models in biotech.* Nature Biotechnology, 2000. 18(Supplement): pp. IT3–IT4.

194. James, R., *Differentiating genomics companies.* Nature Biotechnology, 2000. 18(2): pp. 153–55.

195. Bassett, P., *Paring down the company: The effect of consolidation in the European biotech sector.* Nature Biotechnology, 2003. 21(7): pp. 829–30.

196. Thiel, K.A., *Goodbye Columbus! New NRDOs forgo discovery.* Nature Biotechnology, 2004. 22(9): pp. 1087–92.

197. Dean, P.M., E.D. Zanders, and D.S. Bailey, *Industrial-scale, genomics-based drug design and discovery.* Trends in Biotechnology, 2001. 19(8): pp. 288–92.

198. Beeley, L.J. and D.M. Duckworth, *The impact of genomics on drug design.* Drug Discovery Technology, 1996. 1(11): pp. 474–80.

199. Pisano, G.P., *Searching for viable biobusiness models.* Genetic engineering news, 2007. 27(8): pp. 6–8.

200. Bains, W., *The Timetable of innovation.* Drug Discovery World, 2006. 2006(2): pp. 3–12.

201. Sheridan, C., *Diversa restructures, raising questions over bioprospecting.* Nature Biotechnology, 2006. 24(3): p. 229.

202. Silverman, E., *Who took the tech out of biotech.* Nature Biotechnology, 2006. 24(3): pp. 255–6.

203. Morrow, K.J., *Guaging nanobiotech's promise realistically.* Genetic engineering news, 2006. 26(21): pp. 1–3.

204. Dorey, E., *Skye pharma chair goes,* 2006 Chemistry and Industry, 6 February 2006. p. 10.

205. Stone, M., *Biotech venture capital examines itself.* Nature Biotechnology, 1985. 3(2): pp. 110–13.

206. Jacobs, T., *Bugs or drugs, tortoises or hares.* Nature Biotechnology, 2005. 23(3): pp. 293.

207. Schafer, D.P., *In-licensing as a business model.* Nature Biotechnology, 2004. 20(Bioentrepreneur supplement): pp. BE36–BE9.

208. Hodgson, J., *Validating a business model.* Nature Biotechnology, 2000. 18(4): p. 378.

209. Anon (editorial), *Hybrid rigor mortis.* Nature Biotechnology, 2003. 21(6): p. 585.

210. Sedlak, B.J., *Tool companies shift focus to pharma research.* Genetic engineering news, 2002. 22(14): pp. 18–19.

211. Ratner, M., *As products enter the clinic . . . scientists shown the door.* Nature Biotechnology, 2004. 22(9): p. 1059.

212. Ransom, J., *Niche indications could drive higher valuation.* Nature Biotechnology, 2006. 24(12): p. 1457.

213. Ratner, M., *Syrrx acquisition signams maturation of structure-based discovery.* Nature Biotechnology, 2005. 23(4): p. 400.

214. Fletcher, L., *Genomics companies shop around for chemical expertise.* Nature Biotechnology, 2004. 22(2): pp. 137–8.

215. Petit-Zeman, S., *Regenerative medicine.* Nature Biotechnology, 2001. 19(3): pp. 201–6.

216. Gilson, R.J. and D.M. Schizer, *Venture capital structure: A tax explanation for convertible preferred stock.* Harvard Law Review, 2003. 116: pp. 875–916.
217. Berglof, E., *A control theory of venture capital finance.* Journal of Law, Economics & Organization, 1994. 10: pp. 247–67.
218. Cumming, D., *Agency costs, institutions, learning and taxation in venture capital contracting.* Journal of Business Venturing, 2005. 20(6): pp. 573–622.
219. Mirrlees, J.A., *The theory of moral hazard and unobservable behaviour: Part I.* Review of Economic Studies, 1999. 66: pp. 3–21.
220. Holmström, B., *Moral hazard and observability.* Bell Journal of Economics, 1979. 10(1): pp. 74–91.
221. Trester, J.J., *Venture capital contracting under asymmetric information.* Journal of Banking and Finance, 1998. 22(6–8): pp. 675–99.
222. Altman, L.S., *Term sheet trends in the venture capital market* http://www.nature.com/cgi-taf/gateway.taf?g=6&file=/bioent/building/financing/122003/full/bioent785.html (produced by Nature Biotechnology Entrepreneurship, accessed 12/2/2005).
223. Stein, J.C., *Convertible bonds as backdoor equity financing.* Journal of Financial Economics, 1992. 15: pp. 3–29.
224. Shirley, P., *Venture Capital.* Fiscal Studies, 1994. 15(2): pp. 90–104.
225. Christoffersen, R., *How to improve Series B-round valuations.* Genetic engineering news, 2005. 25(19): pp. 22–5.
226. Cookson, C., *Intercytex sees promise in wound healing, Financial Times* (London), 8 August 2005, p. 22.
227. Friedli, D., I. Dey, and G. Dixon, *That sinking feeling,* Scotland on Sunday, (Edinburgh), 1 August 2004, p. 5.
228. Murray-West, R., *Cyclacel pulls plug on float, Daily Telegraph,* (London), 27 July 2004, p. 27.
229. Firn, D., *Microscience calls of plans for AIM flotation, Financial Times,* (London), 10 July 2004, p. 2.
230. Foley, S., *Business analysis: Why the markets have fallen out of love with biotech companies, Independent,* (London), 29th June 2005, p. 59.
231. Stout, K., *Drivers of M&A: Implications for the UK Biotechnology industry,* in *Tanaka Business School.* 2007, Imperial College: London.
232. Anon, *Pride comes before a fall (and a bloodbath).* Nature Biotechnology, 2002. 20(12): p. 1173.
233. Mitchell, P., *UK Bidding war amid merger mania.* Nature Biotechnology, 2003. 21(4): pp. 343–4.
234. van Brunt, J., *Mergers were hot in 2005* www.signalsmag.com (produced by Recombinant Capital, accessed 22 May 2006).
235. Mitchell, P., *UK Industry consolidation is slow despite big merger.* Nature Biotechnology, 2003. 21(3): p. 215.
236. Lang, L.H.P. and R.M. Stulz, *Tobin's q, corporate diversification, and firm performance.* Journal of Political Economy, 1994. 102: pp. 1248–80.
237. Berger, P.G. and E. Ofek, *Diversification's effect on firm value.* Journal of Financial Economics, 1995. 37: pp. 39–65.
238. Comment, R. and G.A. Jarrell, *Corporate focus and stock returns.* Journal of Financial Economics, 1995. 37: pp. 67–87.
239. Servaes, H., *The value of diversification during the conglomerate merger wave.* Journal of Finance, 1996. 51: pp. 1201–25.

240. John, K. and E. Ofek, *Asset sales and increase in focus.* Journal of Financial Economics, 1995. 37: pp. 105–26.
241. Aggarwal, R.K. and A.A. Samwick, *Why do managers diversify their firms? Agency reconsidered.* Journal of Finance, 2003. 58(1): pp. 71–118.
242. Mansi, S.A. and D.M. Reeb, *Corporate diversification: What gets discounted?* Journal of Finance, 2002. 57(5): pp. 2167–84.
243. Ruffulo, R. *Strategy for re-engineering dsicovery and development to address R&D productivity* in *Drug Discovery Technology Europe 2005.* 2005. London, UK: Informa Ltd.
244. Jacobs, T., *Does biotech M&A benefit investors?* Nature Biotechnology, 2004. 22(9): p. 1085.
245. Meeks, G., *Disappointing marriage: A study of the gains from merger. Occasional Paper 51, Department of Applied Economics, Cambridge.* 1977, University of Cambridge, Department of Applied Economics (Cambridge, UK, 1977).
246. Maybeck, V., *Motives and Mergers as drivers for change in the business model,* in *Institute of Biotechnology.* 2005, Cambridge: Cambridge, UK.
247. Maybeck, V. and W. Bains, *Small company mergers – a good idea for whom?* Nature Biotechnology, 2006. 24(11): pp. 1343–8.
248. Esposito, R.S. and M.J. Ostro, *Strategic consolidation: The biotechnology business model for the 21st century.* Nature Biotechnology, 1999. 17(Supplement): pp. BE16–BE17.
249. Colyer, A., *Financing as a driver for mergers and acquisitions.* Nature Biotechnology, 1999. 17(Supplement): p. BE13.
250. Amihud, Y. and B. Lev, *Risk reduction as a managerial motive for conglomerate mergers.* The Bell Journal of Economics, 1981. 12(2): pp. 605–17.
251. Åstebro, T., *Basic statistics on the success rate and profits for independent inventors.* Entrepreneurship Theory and Practice, 1998. 23(2): pp. 41–8.
252. Jensen, D.G., *Novel model for biotech cluster development.* Genetic engineering news, 2004. 24(16): pp. 1, 14–15.
253. Lindelhöf, P. and H. Löfsten, *Growth, management and financing of new technology-based firms – assess value-added contributions of firms located on and off Science Parks.* Omega, 2002. 30: pp. 143–54.
254. Forbes, D.P., *Are some entrepreneurs more overconfident than others?* Journal of Business Venturing, 2005. 20: pp. 623–40.
255. Graham, P., *How to start a start-up* http://www.paulgraham.com/start.html (produced by Paul Graham, accessed 24 March 2005).
256. Andrews, A.O. and T.M. Welbourne, *The people/performance balance in IPO firms: The effect of the chief executive officer's financial orientation.* Entrepreneurship Theory and Practice, 2000. 25(1): pp. 93–106.
257. Birley, S. and P. Westhead, *A comparison of new businesses established by 'novice' and 'habitual' founders in Great Britain.* International Small Business Journal, 1993. 12(1): pp. 38–57.
258. Westhead, P. and M. Wright, *Novice, portfolio and serial founders: Are they different?* Journal of Business Venturing, 1998. 13: pp. 173–204.
259. Busenitz, L.W., J.O. Fiet, and D.D. Moesel, *Signalling in venture capitalist-new venture team funding decisions: Does it indicate long-term venture outcomes?* International Small Business Journal, 2005. 29(1): pp. 1–12.
260. North, J., J. Curran, and R. Blackburn, *Quality and small firms: A policy mismatch and its impact on small enterprises,* in *Small firms: Entrepreneurship in the*

nineties, D. Deakins, P. Jennings, and C. Mason, eds 1997, Paul Chapman Publishing: London. pp. 112–26.

261. Sun, H. and T. -K. Cheng, *Comparing reasons, practices and effects of ISO9000 certification and TQM implementation in Norwegian SMEs and large firms.* International Small Business Journal, 2002. 20(4): pp. 421–42.

262. Storey, D.J., *Exploring the link, among small firms, between management training and firm performance: A comparison between the UK and other OECD countries.* International Journal of Human Resource Management, 2004. 15(1): pp. 112–30.

263. Elton, E.J., M.J. Gruber, and C.R. Blake, *Incentive Fees and Mutual Funds.* Journal of Finance, 2003. 58(2): pp. 779–804.

264. Cumming, D., G. Fleming, and J.-A. Suchard, *Venture capitalist value-added activities, fundraising and drawdowns.* Journal of Banking and Finance, 2005. 29(2): pp. 295–331.

265. Anon, *50 million dollars for leap into bio-tech, Times,* (London), 24th May 1987, p. 3.

266. Fallon, I., *Bio-raidable, Sunday Times,* (London), 1 July 1990, p. 1.

267. Chong, L., *Goldman staff share $16bn in bonus bonanza, Times,* (London), 12 December 2006, p. 1.

268. Bawden, T., *Lehman and Bear Stearns join banking bonus boom, Times,* (London), 15 December 2006, p. 3.

269. Ang, J., B. Lauterbach, and B.Z. Schreiber, *Pay at the executive suite: How do US banks compensate their top management teams?* Journal of Banking and Finance, 2002. 26(6): pp. 1143–63.

270. Balboa, M. and J. Martí, *Factors that determine the reputation of private equity managers in developing markets.* Journal of Business Venturing, 2007. 22(4): pp. 453–480.

271. Stewart, T.A., *Corporate social responsibility: Getting the logic right.* Harvard Business Review, 2006. 2006(December): p. 14.

272. Frank, L., T. Tschernine, and F. Schöhharting, *Venture capital with a European twist.* Nature Biotechnology, 1999. 17(Supplement): pp. BE23–24.

273. Ross, S.A., *The economic theory of agency: The Principal's problem.* American Economic Review, papers and proceedings, 1973. 63(2): pp. 134–9.

274. Jeng, L.A. and P.C. Wells, *The determinants of venture capital funding: Evidence across countries.* Journal of Corporate Finance, 2000. 6: pp. 241–89.

275. Koberstein, W. and K. Robinson, *The secure IPO launch plan,* in *BioExecutive International.* 2005. pp. 59–67.

276. Bergemann, D. and U. Hege, *Venture financing, moral hazard, and learning.* Journal of Banking and Finance, 1998. 22(6–8): pp. 703–35.

277. Silverman, E., *The hermit crab solution.* Nature Biotechnology, 2006. 24(3): pp. 245–7.

278. Gompers, P.A., *Grandstanding in the venture capital industry.* Journal of Financial Economics, 1996. 42: pp. 133–56.

279. Lee, P.M. and S. Wahal, *Grandstanding, certification and the underpricing of venture capital backed IPOs.* Journal of Financial Economics, 2004. 73: pp. 375–407.

280. Neus, W. and U. Walz, *Exit timing of venture capitalists in the course of an initial public offering.* Journal of Financial Intermediation, 2005. 14: pp. 253–77.

281. Durman, P., *Biotechs 'not ready' to float, Sunday Times,* (London), 23 May 2004, p. 3.

282. Wang, C.K., K. Wang, and Q. Lu, *Effects of venture capitalists' participation in listed companies*. Journal of Banking and Finance, 2003. 27(10): pp. 2015–34.
283. Jain, B. and O. Kini, *Venture capitalists participation and the post-issue operating performance of IPO firms*. Managerial and Decision Economics, 1995. 16: pp. 593–606.
284. Warren, W.J., *University start-up licensing tactics*. Genetic engineering news, 2004. 24(21): pp. 20–2.
285. Parker, R., *The myth of the entrepreneurial economy: Employment, innovation and small firms*. Work Employment and Society, 2002. 15(2): pp. 373–84.
286. Lambert, R., *Lambert Review of Business-University Collaboration: Final report*. 2003, HMSO (London, UK).
287. Storey, D.J. and S. Johnson, *Are small firms the answer to unemployment?* 1987, Employment Institute (London, 1987).
288. Birch, D., *The job creation process*. 1979, M.I.T. programme on neighborhood and regional change (Cambridge, MA, 1979).
289. Marlow, S., *Commentary on Parker*. International Small Business Journal, 2002. 20: p. 111–2.
290. Ross, H., *Supporting High Technology start-ups: The Scottish experience*, in *New technology-based firms in the new millennium*, W. During, R. Oakey, and M. Kipling, eds 2000, Permagon: Oxford. pp. 8–15.
291. Michael, A., *Financing doubts surround hundreds of German companies*. Nature Biotechnology, 1997. 15(12): pp. 1334–5.
292. Fürst, I., *Concern mounts over Munich's BioRegio money*. Nature Biotechnology, 1998. 16(7): pp. 614–15.
293. Fürst, I., *Boom to doom for German biotech?* Nature Biotechnology, 1998. 16(11): p. 997.
294. Turok, I., *Disconnected measures: European structural fund support for small business*, in *Small Firms: Entrepreneurship in the Nineties*, D. Deakins and P. Jennings, eds 1998, Paul Chapmen Publishing: London. pp. 141–61.
295. Giles, J., *Tax change curtails UK University spin-offs*. Nature, 2004. 429: p. 6.
296. Cullen, K., *UNICO – measuring change and making a difference in the UK*, in *From IP to IPO: Key issues in commercialising university technology (Supplement to Intellectual Property Management magazine)*, J. Wild, ed. 2005, Intellectual Property Management magazine: London. pp. 11–12.
297. Cumming, D., *The structure, governance and performance of UK Venture Capital Trusts*. Journal of Corporate Law Studies, 2003. 3(2): pp. 401–27.
298. DTI, *Trade and Industry – Thirteenth Report*. 2003, Department of Trade and Industry (London, 9 May 2003).
299. OST, *The University Challenge Fund guidelines – outline application stage*. 1998, Office of Science and Technology, Department of Trade and Industry (London, 1/3/1998).
300. Hewitt, P., *Answer to parliamentary question*. 2003, Hansard: London. p. 331W.
301. Johnstone, H., *Equity gaps in depleted communities*, in *New technology-based firms in the new millennium*, W. During, R. Oakey, and M. Kipling, eds 2000, Permagon: Oxford. pp. 84–94.
302. Bottazzi, L. and M. Da Rin, *Venture capital in Europe and the financing of innovative companies*. Economic Policy, 2002. 17(34): pp. 229–70.
303. Lynn, J. and A. Jay, *Yes, Prime Minister*, vol II, pp 130–1. 1987, London: BBC Books.

304. Anon, *Blair calls for 'big push' for UK biotech industry*, in *SCRIP World Pharmaceutical News*. 2007. p. 3.
305. Aernoudt, R., *European policy towards venture catital: Myth or reality?* Venture Capital, 1999. 1(1): pp. 47–58.
306. Enserink, M., *France hatches 67 California wannabes*. Science, 2005. 309(5734): p. 547.
307. NAPF, *Institutional Investment in the UK Six Years On*. 2007, National Association of Pension Funds (London, January 2007).
308. Washburn, J., *University Inc. The corporate corruption of higher education*. 2004.
309. Patten, C., *Europe pays the price for spending less*. Nature, 2006. 441(7094): pp. 691–3.
310. DeFrancesco, L., *Company founders: Voices of experience* http://www.nature.com/nbt (produced by Nature Biotechnology, accessed 1 April 2005).
311. Hollon, T., *Aiming for the next winner* http://www.signalsmag.com (produced by Signals Magazine, accessed 1 April 2005).
312. Bains, W., *Fraud and Scandal in biotech*. Nature Biotechnology, 2006. 24(7): pp. 745–7.
313. Evans, C., *Building value in European biotech*. 2002, IIR Conferences – presented material: IIR Biotechnology Conference, 19–20/11/2002, ExCeL Centre, London, UK.
314. Wong, A., *Angel finance: The other venture capital (University of Chicago working paper)*. 2002: University of Chicago.
315. J. Robert Scott, WilmerHale, and Ernst & Young LLP, *Compensation and Entrepreneurship Report in Life Sciences* http://www.compstudy.com/LFdefault.aspx (produced by Harvard Business School, accessed 20 August 2006).
316. Feldbaum, C.B., *A bill of rights for bioentrepreneurs*. Nature Biotechnology, 1998. 16(Supplement): p. 18.

Company Index

Note an *italicised* entry means a mention in a figure or table, other entries are references in the text.

3i 36
AbCam 6, 188
Abgenix *78*
Acambis 30, 60
Actelion *60*
Advent 26, 27
Affymetrix 54, 181
Air Products 27
Alizyme *107*, 196
Amedis Pharmaceuticals 100, 137
Amgen 20, 25, 54, *60*, 177
  and EPO 6
Amylin 28
AquaPharm Biodiscovery 59
Arakis 132–3
Ardana Bioscience *107*
Argenta 119
Ark Therapeutics 13, *107*
Arrow Therapeutics 131
Artemis Inc 180
Athena Neurosciences 70
Atugen 180
Aviron 70
Avlar Bioventures 9, 13

Basilea *60*
BIL see Biotechnology Investments
  Ltd.
Biofocus *107*, 119
Biofusion 75
Biogen 13, 20, 25–8, 37, *60*
  history 25
Biotechnology Investments Ltd. 26,
  29, 46, 154
Biovex 28
Biovitrum 85
British and Commonwealth Shipping
  27
British Biotechnology Ltd. 20, 182
British Technology Group 27, *60*

Cambridge Antibody Technology 7,
  60, 78
Cambridge Life Sciences 26
Camitro 63
Cantab Pharmaceuticals 28
Celera 82, 3
Celgene 20, 84
Cell Genesys 84
Celltech 5, 20, 26–7, 85, 182
Cetus 5
Charterhouse Japet 120
Chiron 5, 25
Chiroscience 13, 18, 121
Ciba Geigy 27
Cogent 26
CV Therapeutics 70
Cyclacel 28, 130–1
Cytomyx *107*

Darwin Moleculer 121
DNAX 47
Domantis 76

EIF see European Investment Fund
Eli Lilly 5, 27
Entelos 63
Ethical Pharmaceuticals 28
European Investment Fund (EIF) 13,
  185–6
Evolutec *107*
Exelesis 78,180

Gene Logic 78, 84
Genentech 5, 9, 20, 25, 54, 60, 177
  Genentech capital used 115
  Genentech IPO 75
  Genentech lawsuits 64
Genetics Institute 78
GenMab 85
Genzyme 20, 25, *60*

Geron   115–6
Gilead   20, *60*
GIMV   36
Grand Metropolitan   27

Henderson Moreley   *107*
Human Genome Sciences Inc   122

IDM   28
Incyte   78
Innogenetics   *60*
Integrated Genetics   *78*
Intercytex   129
IP Group   75, 83

KuDoS   76

Lexicon Genetics   *78*
Lynx Therapeutics   *78*

Maxim Pharmaceuticals   84
Maxygen   84
Medarex   84
Medimmune   25
Medisense   76
Mercator Genetics   *78*
Merlin Biosciences   9, 13, 40
Merlin Ventures   9
Micromet   28
Microscience   131
Millennium Pharmaceuticals   25, 78
Monsanto   27
Morphochem   119
Morphosys   60
Myriad   *78*

Nereus Pharmaceuticals   59
NiCox   85
Nordic Bone   28
Novo Nordisk   5, 17, 182

Onyx Pharmaceuticals   70
Oxford Ancestors   6
Oxford Biomedica   *107*
Oxford Glycosciences   *78*

Paradigm Genetics   *78*
Paradigm Therapeutics   137
Peptech   7
Peptide therapeutics   *60*
Pfizer patents   60
Pharmaceutical Proteins   28
Pharmagene   *107*
Prutech   26

ReNeuron   100, 107, 115–6
Ribozyme Pharmaceuticals   180
Ricera   119

Schering Plough   27
Scotgen   28
Sequana Therapeutics   *78*
Sequenom   38
Serono   17, 25, *60*, 182
   Serono History   5
Skye Pharma   120
Small Molecule Therapeutics Inc   119
Soffinova   94
Solexa   13, 76
Sosei   132–3
Standard Oil   27
Stem Cell Sciences   115
Symyx   70
Synairgen   196

Technical Development Capital
   Ltd   26
Technostart   9
Tegenero   165
Tularik   70

UCB-Celltech see also Celltech   25

VasTox   167, 196
Vectura   107
Vernalis   20, 30, 60, 134
Versant   70
Vysis   78

Xcyte   28, 130
Xcyte   130

Topical Index

Note that as this book is about VC and biotechnology, I have not indexed references to VC or biotechnology.

Academic: making money from
    biotech   38–40
Acquisition   12, 165
  and share structure   130
  valuations   76
Analysts (stock market)   84–5
Angel
  See Business Angel
Anti-dilution provisions   128

BioRegio (Germany)   179
Biotechnology industry   5
Biotechnology Means Business
    initiative   179
Board of directors role   91 ff, 188
  VC directors   97
British Venture Capital Association
  (BVCA)   33
business angels   7–8, 194
  directorships   101
  equity gap   68–9
  inventor productivity   140
Business models   43, 109ff
  and mergers   135–6
  change in   120
  of VC   150
Business quality   46

Capitalization   15–6, 19–20
  at IPO   74
Carried Interest   10–1, 151 ff
  optimal   190
Clusters   141
Company failure   165
  causes   63–4
Company formation   14–5, 24
Company funding (US vs EU)   56
Company origins   52
Consultants   95
Corporate venturing   27, 167

Deal flow   46–7, 66
Debt funding   125
Directors of companies   91
  See also Board of Directors
Drug approvals   17–8
Drug discovery   112 ff
  technology   117
Due diligence   9, 67, 98
Duration of investment process   65

Economic benefit from biotech
    174 ff
Employment   16–7, 176 ff
Entrepreneur   36, 71, 87, 145–6
  impact of VC model   192
  lack of success in VC   159
  removed from companies   102 ff
  reward   37
  support from VC   93
  US s Europe   42
Equity gap   68–70, 180, 183
European lag   14, 23
European Venture Capital Association
  (EVCA)   26
  valuation guidelines   33, 190
Evaluation of investments   89 ff
Exit see also IPO, acquisition 12,
    33–4, 74
  and share structure   129
  and VC grandstanding   166

Facilities cost   62
Failure causes   63–4
  rates   44
FIPCO model   114
Follow-on financing   81 ff
Founders (of companies) See also
    Entrepreneur
  as management   50
  removed from companies   102 ff

Fund life (duration)   10–1, 13, 151
Funding basic science   53–4
   companies   56–7
   through IPO   75
   Funding gap see equity gap   183
   Fundraising (VC) by VCs   156,
      168–9

General partner   151
Genomics   76, 122
Genomics companies   77–79
   business model   115, 117–8
   stock bubble   83
Governance structure   91
Government support for biotech
      169, 173ff
Grandstanding (by VCs)   166

Human Resources management   148–9

Innovation Act (France)   179
Integrated pharmaceutical company
      model   114
Intellectual property see also patents
      147
   commercialization from
      Universities   177–8
Investment   55 ff
   and equity gap   69–70
   time to make   65
   US vs EU   56–7
Investment evaluation   89 ff
investment process   66–8
investment rounds   57
Investor readiness   70
IPO   12, 73 ff, 165
   and grandstanding   166
   and share structure   129
   as preferred exit   18
   performance and management   106
ISO 9000   149

Legal actions of biotechs   64
Licensing   116
Lifestyle business   6
Limited partners   151, 189
   Return of investment   31
Liquidation events   128 ff
   preferences   122 ff

Living dead investments   164
Location (of company) and
      management   48–50, 141

Management   48–50, 141–4
   quality in Europe   47–9
   VC as   92
   VC perception of role   86 ff
   experience   142–5
   succession   102 ff
Management fees (of funds)   10–1,
      151ff, 190
Medicine development   112 ff
Merger   133 ff, 165
   see also acquisition
   barriers to   138
   and share structure   130
Minority shareholder oppression
      188
Mobility of management   48–50
   of scientists   61–2
Monoclonal antibodies   62
Moral Hazard   124 ff, 162

NED (non-executive directors)   101,
      188
No Research, Development Only
      company (NRDO)   121
Number of companies   14–5, 24

Patents   60, 147
Pension funds   189
PIPEs   81
Platform technology companies   77,
      121
PR and VC behaviour   162
Preference shares   123 ff, 193
Principal-Agent theory   124 ff, 162
Private Investment in Public Entities
      (PIPES)   81
Profit share in VC funds see also
      Carried Interest   151 ff

Quality   148

Rate of formation of companies   15,
      24
Return on Investment (ROI)   10, 30,
      74, 152ff, 191

Salaries   61, 142
SBIR programme   54, 181
Science   61
  public funding   53–4
  quality in Europe   51
Science parks   141
Scientist salaries   61
Secondary offering (financing)   81 ff
Seed funding   69
Serial entrepreneurs   146
Share structure   123 ff
Shareholder oppression   188
Shareholder powers   92 ff
Shareholder veto   91 ff
Shareholders' Agreement   92
SMART awards   181
Software in biotech   63
Spin-outs   52, 169ff, 182, 190
  and funding   55
  founder survival   102
Staging of investments   68
Start-ups capitalization   52, 55, 102

Succession of management
  102 ff
Syndicate (investment)   9

Technology management companies
  70, 82
Trade sale (of company)   see
  acquisition
Tranche (of investment)   68

University Challenge Funds (UK)
  179, 184

Valuation   33, 38, 76, 195
  at IPO   74–76
  of research companies   118
  of VC funds   190
VC investment process   8–10, 97
VC partner   98–9
VC support for business   93 ff
Veto powers (shareholder)   91 ff
Visibility Event   162 ff, 171